BrightRED Study Guide

CfE HIGHER

BIOLOGY

Cara Matthews, Angela Grant and Kathleen Ritchie

First published in 2014 by:
Bright Red Publishing Ltd
1 Torphichen Street
Edinburgh
EH3 8HX

Reprinted with corrections in 2016

A CIP record for this book is available from the British Library.

ISBN 978-1-906736-57-6

With thanks to:
PDQ Digital Media Solutions Ltd, Bungay (layout) and Dr Anna Clark (editorial).
Cover design and series book design by Caleb Rutherford – e i d e t i c.

Acknowledgements
Every effort has been made to seek all copyright-holders. If any have been overlooked, then Bright Red Publishing will be delighted to make the necessary arrangements.

Permission has been sought from all relevant copyright holders and Bright Red Publishing are grateful for the use of the following:

Michael A. Mares (CC BY-SA 2.5)[1] (page 23); Le Do/123rf (page 23); Nina Demianenko/123rf (page 23); monticello/123rf (page 23); OxfordSquare/iStock.com (page 23); bedo/iStock.com (page 23); Chiyacat/iStock.com (page 23); vtupinamba/iStock.com (page 23); bernjuer/iStock.com (page 23); mb-fotos/iStock.com (page 48); TMSK/iStock.com (page 49); sihasakprachum/iStock.com (page 49); nanthm/iStock.com (page 50); PinkBadger/iStock.com (page 52); SamCastro/iStock.com (page 52); ktsimage/iStock.com (page 53); claudiodivizia/iStock.com (page 53); EcoPic/iStock.com (page 53); Mikhail_Levit/iStock.com (page 54); czardases/iStock.com (page 57); felinda/iStock.com (page 63); mchudo/iStock.com (page 63); Ockra/iStock.com (page 63); only_fabrizio/iStock.com (page 63); Sergey_Peterman/iStock.com (page 64); dpullman/iStock.com (page 64); alexfiodorov/iStock.com (page 69); Turpentine Creek Wildlife Refuge (page 70); Scotch Mule Association (page 71); SAATEN-UNION (page 72); International Rice Research Institute (IRRI)/Creative Commons (CC BY 2.0)[2] (page 73); InnaFelker/iStock.com (page 73); GTBacchus (CC BY-SA 3.0)[3] (page 74); chamey/iStock.com (page 74); AwakenedEye/iStock.com (page 74); PaulJRobinson/iStock.com (page 75); Fir0002 (CC BY-SA 3.0)[3] (page 75); United States Department of Agriculture (public domain) (page 78); WebSubstance/iStock.com (page 78); ShaunWilkinson/iStock.com (page 79); An image taken from Farm Animal Welfare Council (FAWC) © Crown Copyright. Contains public sector information licensed under the Open Government Licence v2.0 (page 80); artist-unlimited/iStock.com (page 80); Gilles San Martin (CC BY-SA 2.0)[4] (page 82); Arnar/iStock.com (page 83); Richard Ling (CC BY-SA 2.0)[4] (page 83); rodehi/iStock.com (page 84); Corinata (CC BY-SA 3.0)[3] (page 84); Alex Wild (public domain) (page 86); dobrotica/iStock.com (page 86); Editorial12/iStock.com (page 87); Valenice/iStock.com (page 87); CoreyFord/iStock.com (page 89); Global_Pics/iStock.com (page 89).

(CC BY-SA 2.5)[1] http://creativecommons.org/licenses/by-sa/2.5/
(CC BY 2.0)[2]   http://creativecommons.org/licenses/by/2.0/
(CC BY-SA 3.0)[3] http://creativecommons.org/licenses/by-sa/3.0/
(CC BY-SA 2.0)[4] http://creativecommons.org/licenses/by-sa/2.0/

An exam question taken from the Higher Biology 2010 paper, Section A, Question 16 © Scottish Qualifications Authority (n.b. solutions do not emanate from the SQA) (page 61)

Printed and bound in the UK by Charlesworth Press.

CONTENTS

INTRODUCTION

INTRODUCING CFE HIGHER BIOLOGY

COURSE STRUCTURE

The CfE Higher Biology course is divided into three units:

- Unit 1: DNA and the Genome
- Unit 2: Metabolism and Survival
- Unit 3: Sustainability and Interdependence

The CfE Higher Biology Course is assessed as follows:

UNIT ASSESSMENTS

- Each of the three units is assessed within your school using SQA Unit assessments
- Practical Abilities are also assessed internally. You are required to write a report of one of the investigations that you have carried out.

COURSE ASSESSMENTS

To gain the award of the course you must pass all three units as well as the Course Assessment. The Course Assessment is made up of two components with a total of 120 marks. It will be graded A to D which is determined on the basis of the total mark for both components.

Component 1: Question Paper

You will sit an externally assessed written examination consisting of a paper lasting 2 hours 30 minutes. It will be carried out under exam conditions and marked by SQA. This examination has an allocation of 100 marks and is divided into 2 sections.

- **Section 1** is the Objective Test which is worth 20 marks and consists of 20 multiple choice questions.
- **Section 2** is worth 80 marks and will contain restricted and extended response questions. The extended response questions will have a mark allocation of between 6 and 9 marks.

Marks for this written paper will be distributed approximately proportionately across all three Units and the majority of the marks will be allocated for demonstrating and applying knowledge and understanding. The remainder of the marks will be awarded for applying scientific enquiry, analytical thinking and problem solving skills.

Component 2: Assignment

The assignment is worth 20 marks. You will investigate a relevant topic in biology related to one or more of the key areas in the Higher Biology course and communicate your findings. It will require you to demonstrate your application of skills of scientific enquiry and related biological knowledge and understanding.

EXAM HINTS

You do not need to answer the questions in order. Find a question that you can answer easily, so that you settle your nerves.

Timekeeping is important if you are to complete the whole paper. As a general rule, you should be taking just under one and a half minutes per mark. So allowing ten minutes for settling at the start and checking your paper at the end, the timing for each section should be roughly:

contd

- Section 1: Objective Test: 25–30 minutes
- Section 2: Approximately 1 hour 50 minutes

Remember to look at the mark allocation for each question. Extended response questions worth from 6 to 9 marks will require more lengthy answers so remember to allocate sufficient time for these.

THE STRUCTURE AND AIM OF THIS BOOK

There is no short-cut to passing any course at Higher level. To obtain a good pass requires consistent, regular revision over the duration of the course. The aim of this revision book is to help you to achieve this success by providing you with a concise and engaging coverage of the CfE Higher Biology course material. We recommend that you use this book in conjunction with your class notes, to revise each topic area, prepare for Unit assessments and other internal assessments and in your preparation for the final exam.

The book is divided between the three units of the course. Within each section, there is a double-page spread on each of the sub-sections.

Each double-page spread:

- provides the key ideas and concepts of the sub-section in a logical and digestible manner.
- contains 'Internet links' or 'Don't forget' boxes that flag up vital pieces of knowledge that you need to remember and important things that you must be able to do.
- contains a 'Things To Do and Think About" feature which will extend your knowledge and understanding of the subject, and provide additional interest. Sometimes there are questions to help you check your understanding.
- contains a link to an online test to test your knowledge and understanding of each topic.

REVISION TIPS

- Don't leave your revision until the last minute. Make up a revision schedule, giving yourself enough time to revise thoroughly, and stick to it. Be realistic – you should work around your other activities and remember that you do need to take time off to relax away from your books.
- Find somewhere to study that is quiet and uncluttered. You need space to spread out your work.
- Study for short periods (between 30 and 45 minutes) with short breaks in between to keep your level of concentration higher. Go out of the room where you are studying during each break as this will help you to be refreshed and ready for your next study session.
- Read over each sub-topic at a slower pace than you would usually do and ask yourself questions or read it out loud. Make sure that you understand what you have been reading – you only learn what you understand.
- It's often easier to remember facts if you talk about topics with a family member or a friend. So, find a study buddy who can ask you questions about your work.
- Practice makes perfect; do past-paper practice so that the exam format is as familiar as possible. There are only a few ways in which you can be asked the same question and you will see similar questions and diagrams appearing in many past papers. Doing a past paper against the clock will also help you to get your time management right.
- In the run up to the exams, eat plenty of fresh fruit and vegetables to keep your energy levels up, and make sure that you get a good night's sleep so that you are alert throughout the exam.

Good luck, and enjoy!

THE STRUCTURE OF DNA

DNA

Deoxyribonucleic acid (DNA) contains the code to make all of the tens of thousands of proteins in an organism. Proteins, in the form of enzymes, catalyse the reactions to manufacture a complete individual that is correct for its species. A **gene** is a region of DNA that codes for the specific sequence of amino acids that forms a polypeptide chain. The polypeptide chain can be modified and folded to form one of several proteins. This genetic code is passed on through generations when gametes fuse during fertilisation. Thus, the genetic code is inherited.

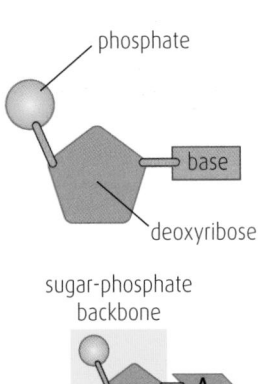
phosphate
base
deoxyribose

DNA IS A DOUBLE-STRANDED HELIX

DNA is made up of sub-units called nucleotides, joined in strands. There are four types of nucleotide, depending on the base: **adenine (A), thymine (T), cytosine (C)** and **guanine (G)**. Each strand is made up of nucleotides which form strong chemical bonds between a phosphate group of one nucleotide and the deoxyribose of another nucleotide: the **sugar–phosphate backbone**. The DNA molecule is **double-stranded** due to the formation of **weak hydrogen bonds** between the bases: adenine always bonds with thymine (**A–T**) and cytosine always bonds with guanine (**C–G**).
The strands are **anti-parallel** meaning they run in opposite directions. Each strand has a 3' end and a 5' end, determined by whether the third or fifth carbon on the sugar molecule of the nucleotide is closest to the end. The 3' end has deoxyribose and the 5' end has phosphate at the end of the strand.

The double-stranded DNA molecule twists to form a **double helix**.

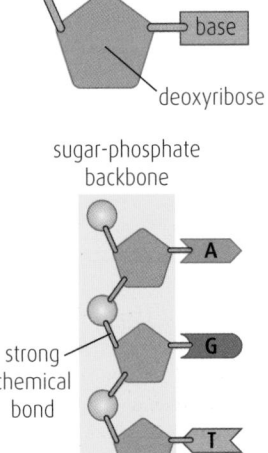
sugar-phosphate backbone
A
G
T
C
strong chemical bond

ORGANISATION OF DNA IN PROKARYOTES AND EUKARYOTES

Organisms fall into two main categories: **prokaryotes** and **eukaryotes**. DNA is found in structures called **chromosomes** or **plasmids**.

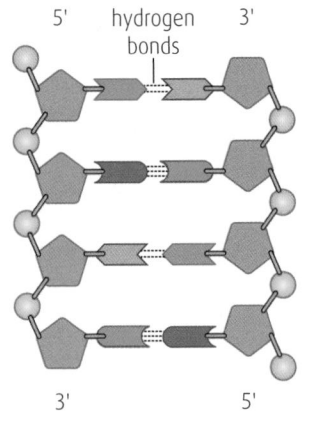
5'   hydrogen bonds   3'
3'   5'

Prokaryotes

Bacteria are examples of prokaryotes. Prokaryotes do not have membrane-bound organelles, such as a nucleus, chloroplasts or mitochondria. Their DNA forms **circular chromosomes** and **plasmids**, which are free in the cytoplasm.

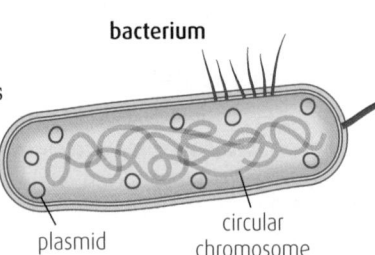
bacterium
plasmid
circular chromosome

Eukaryotes

Plant, animal and yeast cells are all eukaryotes. They are distinguished by having a membrane-bound **nucleus**, along with other membrane-bound organelles such as **mitochondria** or **chloroplasts**.

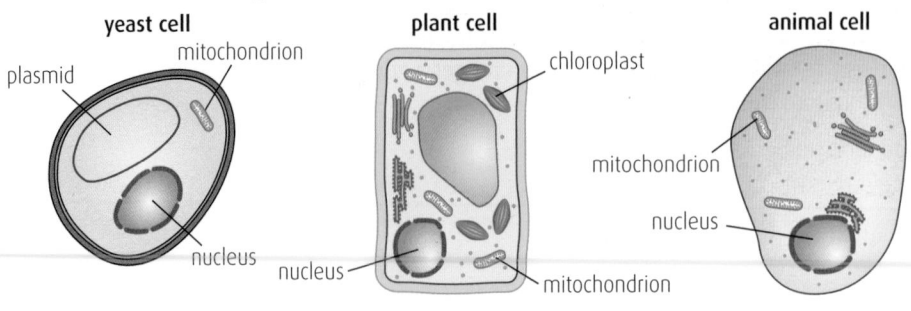
yeast cell
plasmid
mitochondrion
nucleus
plant cell
chloroplast
nucleus
mitochondrion
animal cell
mitochondrion
nucleus

DON'T FORGET

Prokaryotes do not have membrane-bound structures, such as a nucleus, mitochondria or chloroplasts.

ORGANISATION OF DNA IN ORGANELLES

The DNA in eukaryotes is located in the membrane-bound nucleus and takes the form of chromosomes. The DNA in each chromosome is extremely long and thread-like and must therefore be organised into tidy spools (a bit like spools of thread) so it cannot get tangled up with itself and other chromosomal DNA. Each spool is composed of eight proteins and DNA winds around it twice. The resulting 'pearl necklace' is itself coiled tightly to form compact chromosomes. This is how 2 metres of DNA is packed into the microscopic nucleus of every cell in the human body.

Chloroplasts and **mitochondria** have their own **circular chromosomal DNA**.

Mitochondria are located in the cytoplasm of cells, including egg cells. So, they form a component of the zygote at fertilisation. Our mitochondrial DNA is, therefore, inherited exclusively from our mothers. The mitochondrial genes code for essential respiratory enzymes; mutations can be lethal.

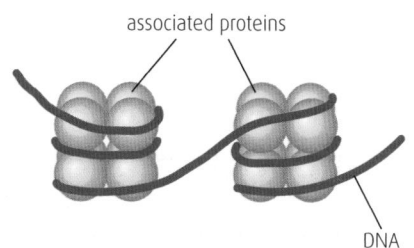
associated proteins / DNA

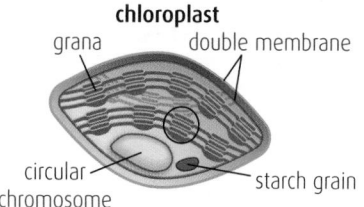
chloroplast — grana, double membrane, circular chromosome, starch grain

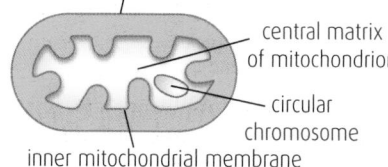
mitochondrion — outer mitochondrial membrane, central matrix of mitochondrion, circular chromosome, inner mitochondrial membrane

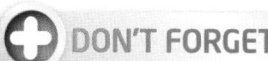

 THINGS TO DO AND THINK ABOUT

1 A molecule of DNA was found to be composed of 32% adenine. Express the ratio of thymine to guanine as a simple whole-number ratio.

2 Label the following on the diagram of a section of DNA:

   a The components of the molecule are represented by numbers 1–4.

   b Show the 3' and 5' ends by adding labels to the black squares.

   c Name the type of bonds labelled A and B.

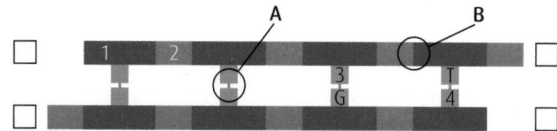

3 Complete the summary table below by adding ticks, if appropriate.

Cell type	Eukaryote	Prokaryote	Nucleus present	Mitochondria present	Chloroplasts present	Plasmids present
Bacterium						
Yeast cell						
Palisade mesophyll cell						
Cheek cell						

4 Complete the summary table below by adding ticks, if appropriate.

Organelle	Thread-like DNA	Circular DNA
Nucleus		
Mitochondrion		
Chloroplast		

COPYING THE CODE

The DNA code has to be copied exactly to make new, fully functioning cells and individuals. This code, while almost identical between members of the same species, has some differences that make it unique to individuals. It is a valuable molecule in crime scene investigation and the development of techniques that allow forensic scientists to make multiple copies for analyses was a major scientific advance.

DNA REPLICATION

DNA must be replicated before cell division can occur, ensuring daughter (new) cells have a complete set of genetic information.

STAGES OF DNA REPLICATION

Requirements for replication:

- DNA
- ATP
- DNA polymerase (enzyme)
- the four types of DNA nucleotide
- primer
- ligase (enzyme)

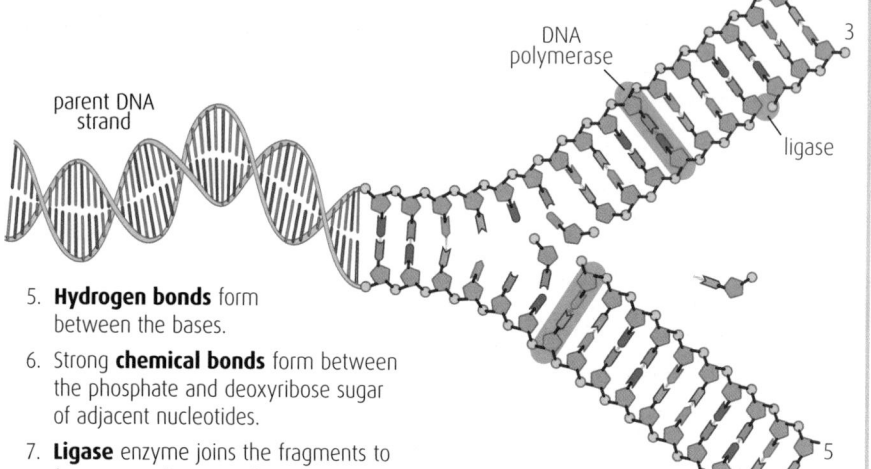

parent DNA strand

DNA polymerase

ligase

1. The DNA molecule **unwinds**
2. Hydrogen bonds break, '**unzipping**' the molecule and exposing the bases of both DNA strands.
3. A **primer** attaches to one end of each exposed DNA template strand.
4. This initiates **DNA polymerase** to add free complementary DNA nucleotides to the **3'** end of the growing strand.

5. **Hydrogen bonds** form between the bases.
6. Strong **chemical bonds** form between the phosphate and deoxyribose sugar of adjacent nucleotides.
7. **Ligase** enzyme joins the fragments to form a complete strand.
8. Each replicated DNA molecule is made of one original template strand and a newly synthesised strand.

DIRECTION OF REPLICATION

DNA polymerase adds nucleotides on to the 3' end of the strand. The strands run in opposite directions, which means that one strand can be built up continuously as the molecule unzips; exposing a 3' end nucleotide – this is the **leading strand**.

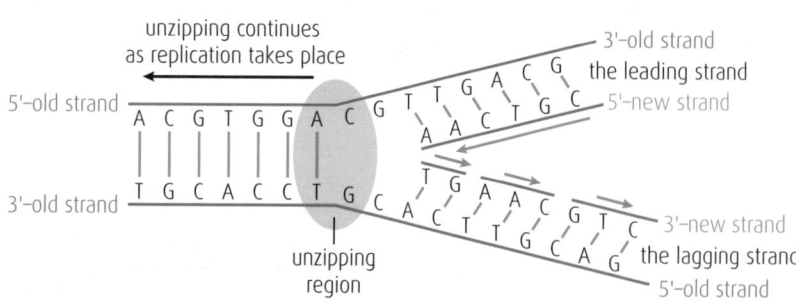

unzipping continues as replication takes place

5'-old strand   A C G T G G A C

3'-old strand   T G C A C C T G

unzipping region

3'-old strand — the leading strand
5'-new strand

3'-new strand — the lagging strand
5'-old strand

However, the other strand 'opens up' the wrong way to add nucleotides to the 3' end. Synthesis using this strand lags behind until enough of the template strand is exposed to add a primer and then add nucleotides to the 3' end. This is repeated in fragments as more template becomes exposed. A complete strand is made when the fragments are joined together by the enzyme **ligase**. This strand is the **lagging strand**.

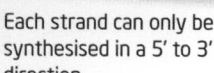

DON'T FORGET

Each strand can only be synthesised in a 5' to 3' direction.

POLYMERASE CHAIN REACTION (PCR)

DNA replication occurs naturally in cells before cell division to make complete copies of DNA. However, specific DNA fragments can be targeted and artificially amplified (increased in quantity) in laboratories by a process known as the polymerase chain reaction.

STAGES OF PCR

Requirements

- an original DNA sample to provide the template
- a stock of four DNA nucleotides
- heat tolerant DNA polymerase (an enzyme)
- a thermal cycler (an automated reaction vessel)
- buffer solution (maintains an optimum pH)
- copies of a DNA primer (which targets the fragment to be amplified)

Heat tolerant DNA polymerase is obtained from bacteria that live around hot springs and are able to withstand high temperatures without their proteins being denatured.

PCR process

- DNA **heated** to 95°C to separate the original DNA strands.
- Sample **cooled** to 55°C. **Complementary primers** can now be added and bind (anneal) to the start of each strand.
- Sample **heated** to 72°C and **Heat tolerant** (thermostable) **DNA polymerase** is added.
- Complementary free DNA nucleotides are added to the **3'** end of the new strands.
- The number of original molecules has now doubled – this is called **amplification**.
- Steps 1–5 are **repeated**, **amplifying** the DNA to make many copies.

Practical Applications of PCR

1 **Forensic science:** a tiny quantity of genetic material is found at a crime scene and it is vital to provide enough material for various tests such as genetic fingerprinting.

2 **Research:** PCR can be used to build up a large bank of stock material from a small initial source e.g. mutated gene sequence. The stock can then be subjected to several types of analyses and be divided up so that several research teams can work on it.

95°C – DNA denatured

55°C – primers anneal

72°C – DNA extension

Taq polymerase

forward primer

reverse primer

3 **Early detection of infection:** PCR can be used to amplify viral DNA when only a few cells out of millions are infected e.g. HIV infection. Otherwise a worried individual may have to wait for months before sufficient cells show infection.

4 **Archaeobiology:** the DNA found in mummified bodies, skeletal remains or insects trapped in amber is usually highly degraded. However any small samples may be amplified and sequenced to look for connections to other mummified remains, find living relations or build a picture of life at that time

5 **Phylogenetics:** this looks at evolution and maps when species split from each other by how much their DNA differs because of mutations. Again small samples from mummified or fossil evidence can be used.

THINGS TO DO AND THINK ABOUT

1 Complete this diagram to show the direction of DNA replication. Show the 3' and 5' ends and draw the leading strand with a continuous line and the lagging strand with a broken line.

2 Explain why it is necessary to increase the temperature to 92°C during a PCR cycle, then cool it to 55°C.

3 How many copies would you have of a single DNA fragment after 10 cycles through a PCR machine?

4 Heat tolerant DNA polymerase is obtained from bacteria that live in the margins of hot springs and are adapted to withstand a temperature range of 50 to 80°C. What would happen if you used DNA polymerase that was obtained from bacteria that live in a temperature range from −5 to 45°C?

GENE EXPRESSION AND PROTEINS

Each protein in an organism has a specific sequence of amino acids and their order in the polypeptide chain is determined by the sequence of bases in the genetic code. The bases (A, T, C and G) that make up the genetic code are the same for all organisms. Differences in the sequence and number of these four bases give rise to a vast number of different proteins. Only a fraction of the genes in a cell are expressed (translated into protein).

DON'T FORGET

All proteins on earth are made from a combination of 20 amino acids which are held together by peptide bonds. Different combinations and numbers of amino acids result in different proteins.

VIDEO LINK

Find out some of the things that proteins do at www.brightredbooks.net

STRUCTURE OF PROTEINS

Once the primary structure has formed, the polypeptide becomes coiled and is held together by hydrogen bonds.

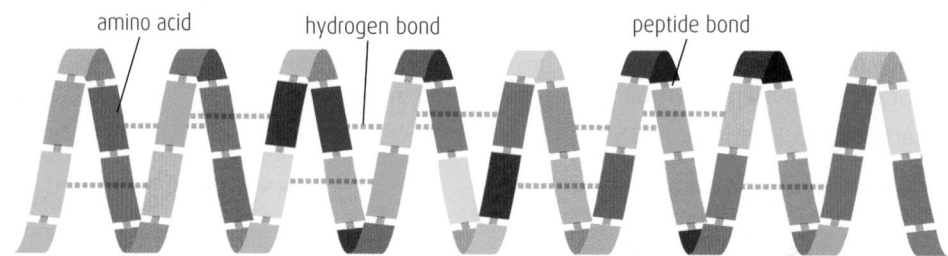

amino acid   hydrogen bond   peptide bond

Finally, the polypeptide forms sheets (fibrous proteins) or is wound up to form a ball (globular proteins). The three-dimensional shape is determined by **hydrogen bonds** that hold the chains together. The protein's shape is due to interactions between amino acids, which cause it to twist and fold in specific ways; hydrophobic areas of the chain move inwards and away from the watery external environment, and positively and negatively charged amino acids move towards each other.

polypeptide   polypeptide

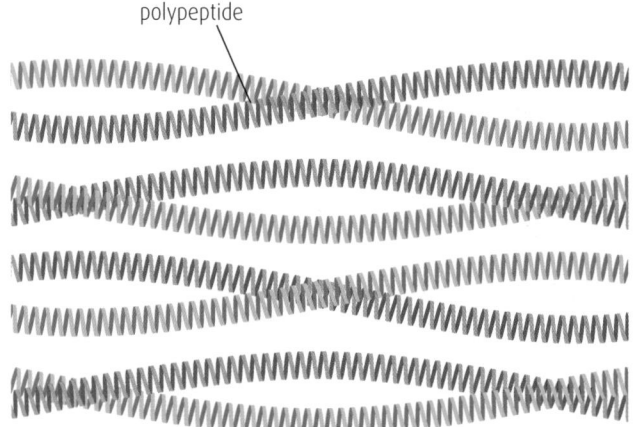

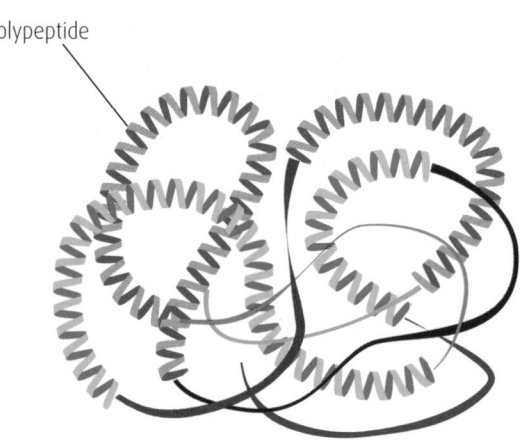

fibrous proteins – flat sheets   globular proteins – wound into a ball

GENE EXPRESSION IN EUKARYOTES

Gene expression determines the phenotype of an organism. The proteins produced (enzymes, hormones, structural and transport proteins) all work together to determine the characteristics that are typical for a species.

A specialised cell, such as a skeletal muscle cell, expresses two types of gene: those that are vital for its maintenance, such as the genes coding for respiratory enzymes, and those that are vital for its specialised function, for example the genes coding for the contractile proteins actin and myosin in muscle cells. All other genes will be inhibited (switched off).

Characteristics that are unique to the individual (such as height and body mass) are the result of genotype and are influenced by **intra-cellular and extra-cellular environmental factors**. Diet, activity levels, stress levels and infection can all affect an organism's internal environment, triggering changes in pH, chemical and hormone production. This, in turn, affects gene expression by switching genes on or off.

THINGS TO DO AND THINK ABOUT

1   Name x, y and z on the diagram of the polypeptide.

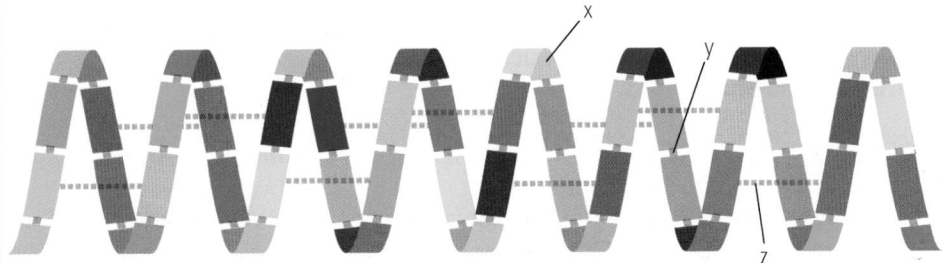

2   What factors determine skin tone?

3   Gel electrophoresis is a technique that is used to separate and identify proteins. Each sample is loaded into a well in a gel. 'Ladders' containing proteins of known molecular weight are loaded at each end for comparison. An electric current is applied which 'pushes' the proteins along the gel; the heaviest proteins travel the least distance and the lighter proteins travel furthest.

A gene mutation occurs in fish which results in a protein with a shorter polypeptide chain being produced. Which sample, A or B, shows the mutation? Justify your answer.

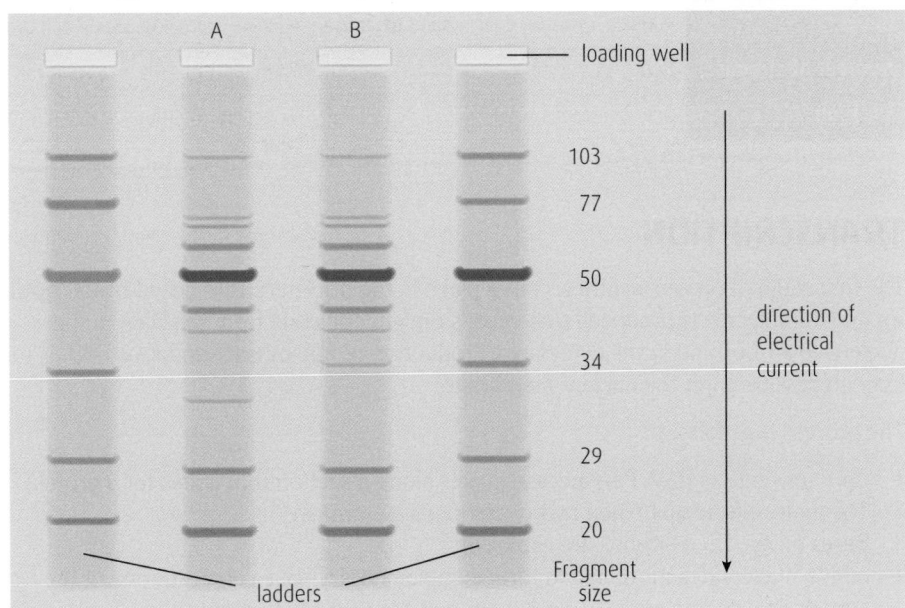

 ONLINE TEST

Once you've learned about this topic, test your knowledge at www.brightredbooks.net

PROTEIN SYNTHESIS 1

The primary structure of a protein is determined by the sequence of nucleotide bases in a DNA strand which cannot leave the nucleus. Proteins are synthesised outside the nucleus on **ribosomes**. A copy of the DNA code is carried as a message from the DNA to the ribosome by another type of nucleic acid called **ribonucleic acid (RNA)**.

STRUCTURE AND FUNCTIONS OF RNA

Ribonucleic acid (RNA) is a single-stranded molecule made of nucleotide sub-units.

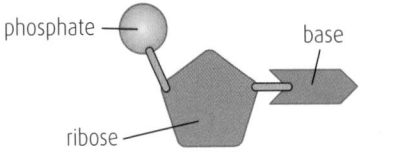 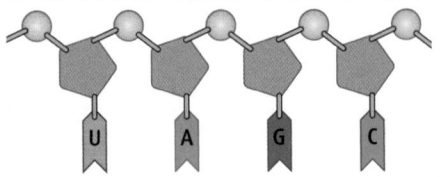

Each nucleotide consists of a phosphate group, ribose sugar and a base. There are four different bases: adenine, **uracil**, guanine and cytosine. Strong chemical bonds form between the phosphate group of one nucleotide and the ribose of the next nucleotide to form an RNA molecule.

You should be familiar with the following types of RNA, all of which are involved in protein synthesis.

- Messenger RNA (mRNA) is formed during transcription of DNA in the nucleus and is the template for protein synthesis at the ribosomes.
- Transfer RNA (tRNA) carries amino acids to the ribosomes for translation of the genetic code.
- Ribosomal RNA (rRNA) binds to proteins to form ribosomes.

COMPARISON OF DNA AND mRNA

	DNA	mRNA
Type of sugar	Deoxyribose	Ribose
Bases	Adenine, cytosine, guanine and **thymine**	Adenine, cytosine guanine and **uracil**
Number of strands	Two	One
Location	Only in nucleus	Moves from nucleus to cytoplasm

TRANSCRIPTION

The first stage of protein synthesis takes place in the nucleus and is called **transcription**. An mRNA molecule is produced that carries the genetic code from the DNA in the nucleus to a ribosome in the cytoplasm. Production of mRNA is essential, as DNA is too large to pass through the nuclear membrane.

The process is as follows:

1. DNA unwinds as **RNA Polymerase** moves along a section that codes for a protein.
2. The molecule 'unzips' when hydrogen bonds are broken.
3. Bases on the DNA strands are exposed.
4. mRNA nucleotides move in and form complementary base pairs with one of the DNA strands (the coding strand). Weak hydrogen bonds form. Cytosine always pairs with guanine; adenine on DNA pairs with uracil on mRNA, and thymine on DNA pairs with adenine on mRNA.

contd

5   Strong chemical bonds form between the phosphate of one nucleotide and the ribose of the next nucleotide, building the mRNA strand.

6   The weak hydrogen bonds that were holding the DNA and mRNA strands together break, allowing the **mature mRNA transcript** to pass out of the nucleus and enter the cytoplasm.

7   Hydrogen bonds reform between the two DNA strands, and the DNA molecule rewinds to form a double helix.

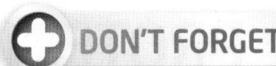

DON'T FORGET

A triplet of bases on mRNA is called a **codon,** and codes for one amino acid.

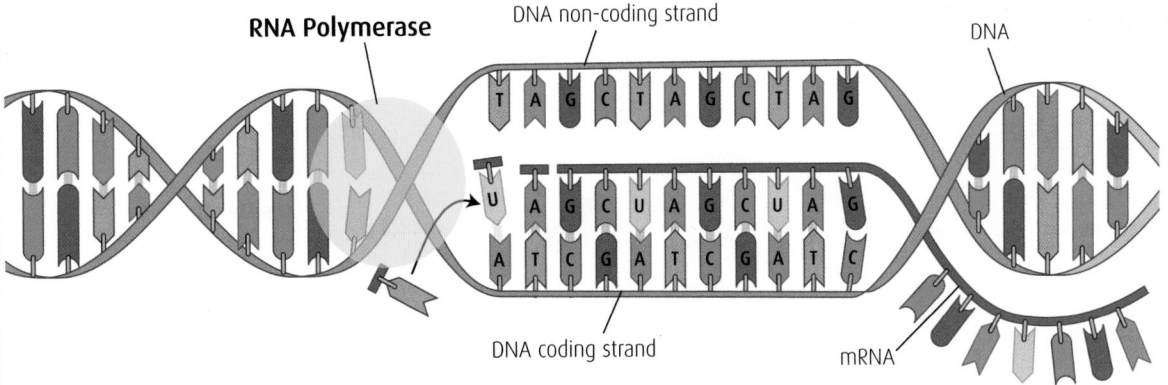

RNA SPLICING

The original DNA template contains the code to make protein. However, it also contains non-coding sections.

The functions of different non-coding regions are as follows:

- to regulate transcription, e.g. by having binding sites for other chemicals that can switch off neighbouring genes
- transcribed to form RNA but not translated into protein
- some regions make no sense and are thought to be the result of mistakes.

The coding regions are called **exons** and the non-coding regions are called **introns**.

| exon | intron | exon | intron | exon | intron | exon | intron |

After the primary transcript has been produced the introns must be cut out.

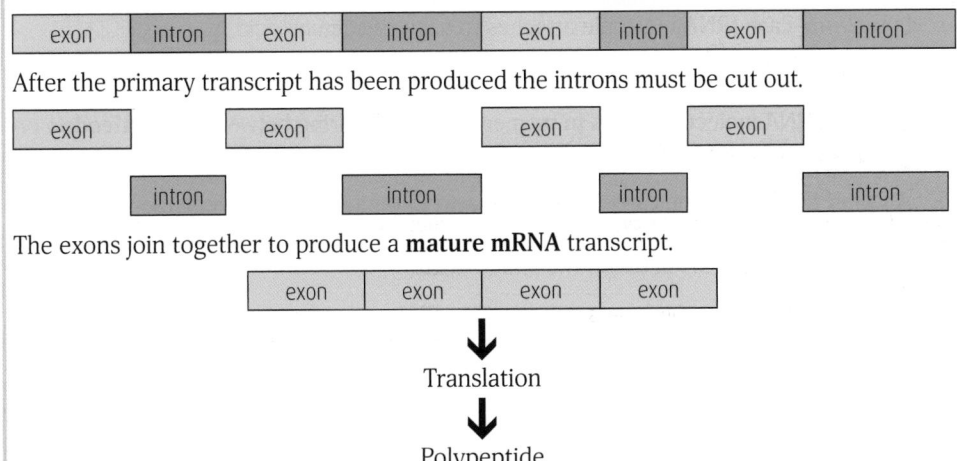

The exons join together to produce a **mature mRNA** transcript.

| exon | exon | exon | exon |

↓
Translation
↓
Polypeptide

THINGS TO DO AND THINK ABOUT

1   Write the complementary mRNA code to this strand of DNA:
    AGGCTAACTGCAATCGAAATG

2   List all the raw materials required for transcription to take place.

3   Describe the sequence of events in transcription.

4   What is the difference between an intron and an exon?

5   How does a primary transcript differ from a mature mRNA strand?

ONLINE TEST

Head online and test yourself on this at www.brightredbooks.net

PROTEIN SYNTHESIS 2

The mRNA molecule formed during transcription leaves the nucleus via a nuclear pore and attaches to a **ribosome**. The ribosome is made up of two units composed of ribosomal RNA (rRNA) and proteins. The nucleic code is now ready to be translated into an amino acid sequence.

DON'T FORGET

Base pairing within **transfer RNA** molecules causes the tRNA to fold into a clover-leaf shape with two distinct regions: an exposed triplet anticodon site and an attachment site for a specific amino acid. The anticodon matches to its complementary triplet codon on the mRNA strand, bringing its specific amino acid with it.

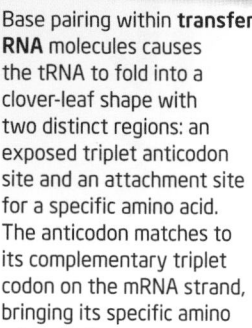

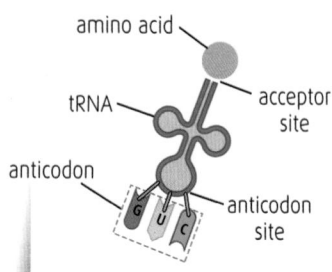

DON'T FORGET

You only need to remember that start and stop codons exist. You do not need to remember the actual codon sequences for individual amino acids.

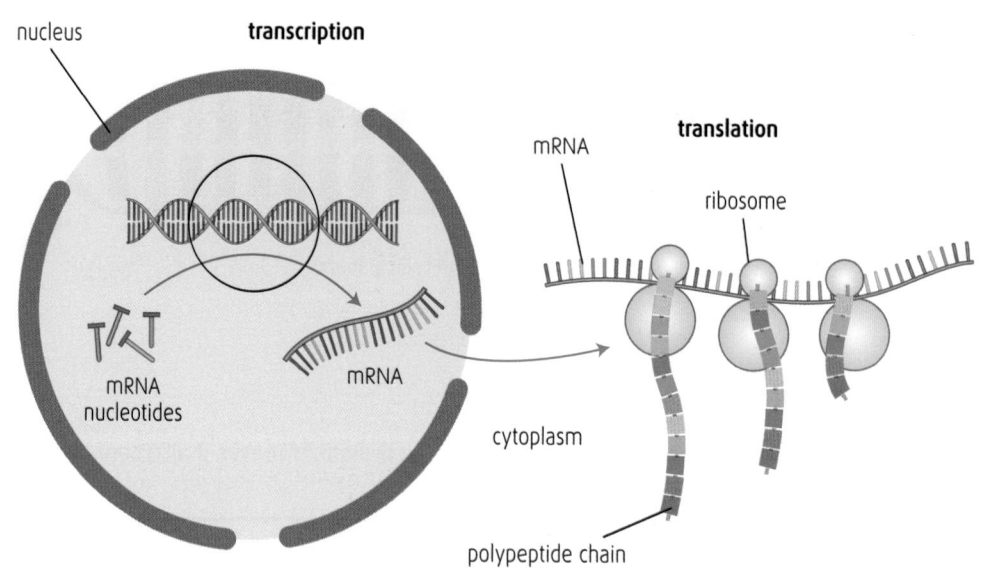

TRANSLATION

1   Translation is initiated by the **start codon** (AUG) on the mRNA strand.

2   Transfer RNA (tRNA) molecules become attached to amino acid molecules in the cytoplasm. Each tRNA molecule attaches to a specific amino acid.

3   tRNA molecules in the cytoplasm transport amino acids to the **ribosome**.

4   The first tRNA molecule moves in by means of base pairing between the **anticodon** on the tRNA molecule and the complementary **codon** on the mRNA strand.

5   Another tRNA molecule carries an amino acid to the ribosome. Complementary pairing between codon and anticodon brings the amino acids in line beside each other. A **peptide bond** forms between the amino acids.

6   The first tRNA molecule detaches from the mRNA and is free to collect another amino acid from the cytoplasm.

7   As translation progresses, the ribosome moves along the mRNA molecule exposing the third codon, allowing a third tRNA molecule to bring a third amino acid into position.

8   This process repeats until the **stop codon** (UGA) is reached at the 3' end of the mRNA strand and the newly formed polypeptide chain is complete.

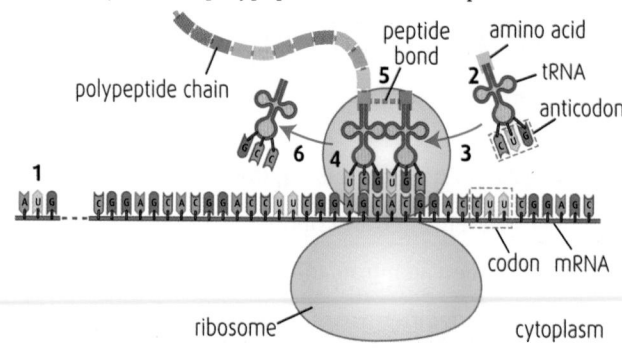

ONE GENE, MANY PROTEINS

After transcription

One gene may code for more than one protein. The codes for different proteins may be made by cutting out alternative regions (**splicing**) of the mRNA transcript.

An **intron** (non-coding region) for one protein may be an **exon** (coding region) for a different protein. Different mRNA molecules are produced from the same primary transcript depending on which RNA segments are treated as exons and which as introns.

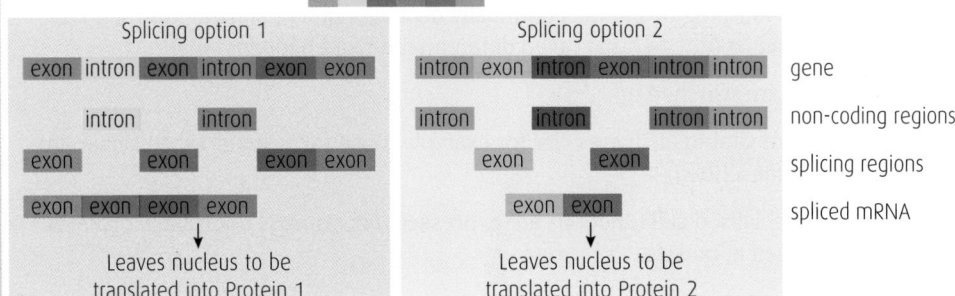

After translation

Different proteins may also be made *after* translation during **post-translational modification**. Polypeptides may be adapted in several ways after transcription:

1   Amino acids may be **cut out** from the polypeptide chain.

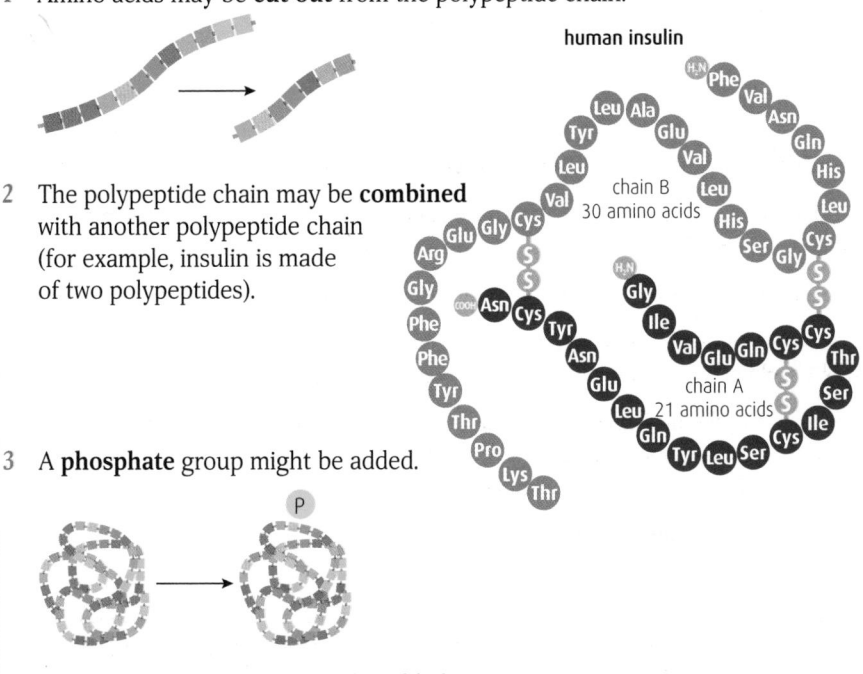

2   The polypeptide chain may be **combined** with another polypeptide chain (for example, insulin is made of two polypeptides).

3   A **phosphate** group might be added.

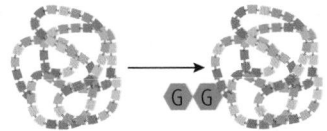

4   A **carbohydrate group** might be added.

VIDEO LINK

Watch an animation of mRNA splicing at www.brightredbooks.net

THINGS TO DO AND THINK ABOUT

1   Circle the correct answer

  DNA polymerase/RNA polymerase builds the mRNA primary transcript.

2   Describe how one mRNA strand can produce several different proteins.

ONLINE TEST

How well have you learned this topic? Take the test at www.brightredbooks.net

DIFFERENTIATION IN MULTICELLULAR ORGANISMS

DIFFERENTIATION

As soon as fertilisation occurs, the zygote (fertilised egg) begins to divide by mitosis. It forms a ball of identical and **unspecialised** cells. Then, as the cells become specialised for specific functions, regions in the continually dividing embryo start to look and act differently from each other. This is called differentiation and happens when:

- **some genes** are switched off

- genes that are vital to all living cells, for example those for respiratory enzymes, are expressed (transcribed)

- genes for specialised cell functions are **expressed**, for example contractile proteins are synthesised in muscle.

Specialised cells can be somatic (body cells) or gametes.

MERISTEMS AND STEM CELLS

Plant meristems

Plants have unspecialised regions called **meristems**. These are the only places where growth can occur through mitosis. The daughter cells are unspecialised and will become differentiated as they move further from the meristems. Such cells are **totipotent** which means they can produce all types of cell.

Apical meristems are located in the tips of roots and shoots. Growth here increases the height of the plant.

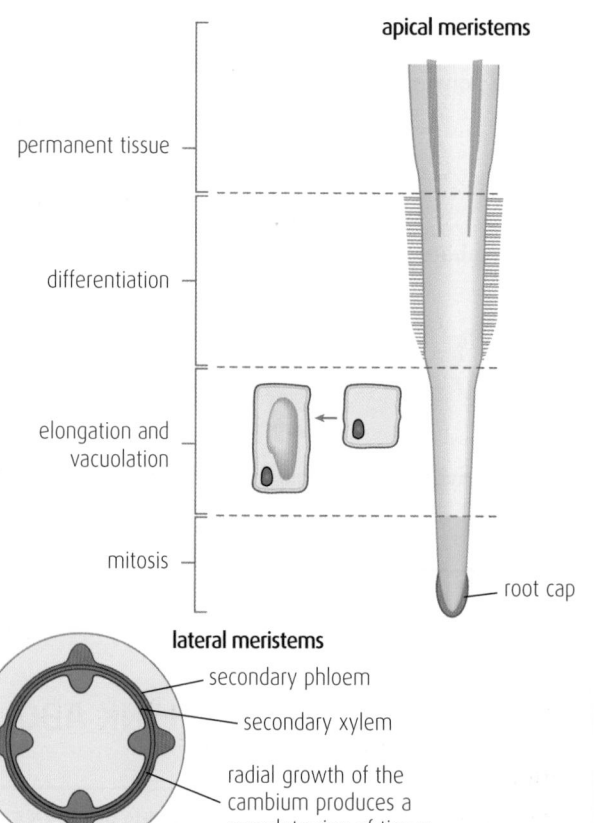

once permanent tissues have been formed, secondary thickening can take place at lateral meristems to increase the thickness of the plant

cells become adapted to form tissues that carry out specialised functions, such as the xylem and phloem

new cells develop vacuoles and elongate, causing the shoot and root to increase in length

undifferentiated cells at the root and shoot-tip divide by mitosis

apical meristems

permanent tissue

differentiation

elongation and vacuolation

mitosis

root cap

Lateral meristems are found in the cambium layer between xylem and phloem tissue. Growth in this region increases the girth of the plant.

lateral meristems

secondary phloem

secondary xylem

radial growth of the cambium produces a complete ring of tissue

contd

Animal stem cells

Animal tissues can grow in all areas.

Stem cells are undifferentiated and can divide to form specialised cells. They are found in all parts of the body and can differentiate into one or a few cell types that are characteristic for their location. The degree of specialisation that occurs depends on when and where cells are found.

Early **embryonic stem cells** are capable of dividing to form *all types* of specialised cell. This is called being **pluripotent**.

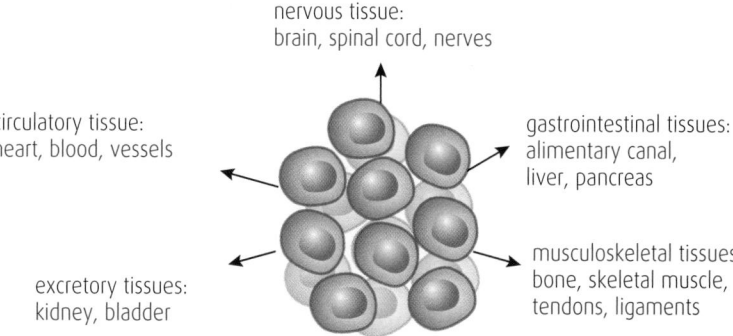

nervous tissue:
brain, spinal cord, nerves

circulatory tissue:
heart, blood, vessels

gastrointestinal tissues:
alimentary canal,
liver, pancreas

excretory tissues:
kidney, bladder

musculoskeletal tissues:
bone, skeletal muscle,
tendons, ligaments

Embryonic stem cells tend to divide into specialised cells, they don't form new stem cells to **renew** (replace) themselves. Embryonic stem cells can, however, be treated in a lab under **in vitro** conditions to renew. This provides a bank of stem cells for research.

Adult stem cells are found in specific locations. They replace specialised cells that have a limited life span, and their ability to differentiate is limited to a few related types. They are **multipotent**, which means they can produce a limited number of cell types.

Some adult stem cells **renew** (divide to make more stem cells) in order to maintain a stock to last the organism's lifetime.

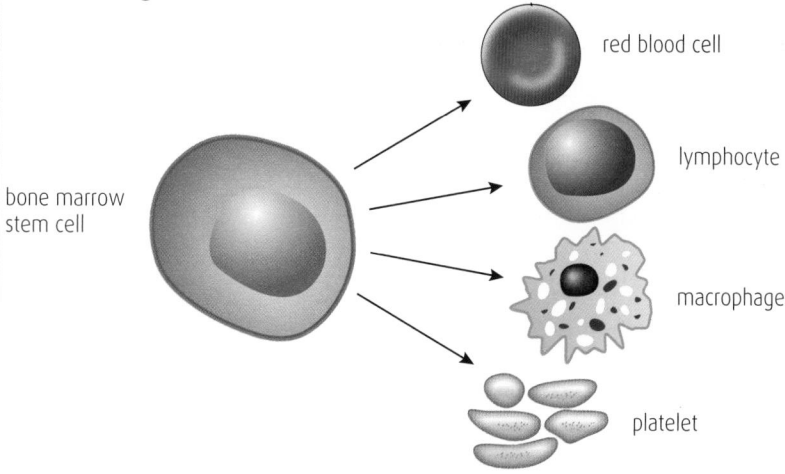

red blood cell

lymphocyte

bone marrow
stem cell

macrophage

platelet

Adult stem cells have yet to be found which can replace every cell type or function, for example produce insulin or differentiate into fully functioning motor neurones.

THINGS TO DO AND THINK ABOUT

1   Why must gardeners ensure that the entire root system is intact, when pulling up weeds?

2   In order to grow a human organ, such as the liver, it would be necessary to have a mixture of different types of adult stem cells. Explain why it is not possible to do this with one type of adult stem cell.

3   What is the role of adult stem cells in the body?

DON'T FORGET

As embryonic cells divide and age, they lose their pluripotency to become either adult specialised cells or multipotent stem cells.

VIDEO LINK

Find out more about the types of stem cells at www.brightredbooks.net

ONLINE TEST

Once you've learned about this topic, test your knowledge at www.brightredbooks.net

RESEARCH AND THERAPEUTIC VALUE OF STEM CELLS

STEM CELL RESEARCH

Stem cell research gives a better understanding of the control of gene expression and differentiation. Stem cells are also used to model cells and tissues to look at the effects of diseases or drug therapy. A limitation of this research is that it can't look at the interaction with other cells, tissues or chemicals in the organism. A key aim is to promote the therapeutic use of stem cells to replace damaged or diseased tissue.

The production of stem cell lines can originate from several sources and is under strict control.

SOURCES OF STEM CELLS

Embryonic Stem Cells

Excess, weaker embryos are donated by couples undergoing IVF treatment → The medical history of donors is thoroughly investigated before embryos are accepted → Embryos are transferred to aseptic lab Embryos are transferred to culture medium and grown at 37°C to the blastocyst stage – day 8 of development → The inner mass of the blastocyst is removed and cultured into pluripotent stem cell lines → Stem cells can be used for research or to replace damaged or diseased tissue

These cells are pluripotent so have the potential to differentiate into every type of cell. Their use in research is important to study the development of the embryo and the causes and treatment of diseases of early embryonic development.

Induced Pluripotent Stem Cells (iPSCs)

In this technique adult cells are reprogrammed to be undifferentiated, self-renewing, pluripotent stem cells.

Reprogramming factors introduced that alters the switching off of genes ↓

Adult, specialised cells e.g. skin cells → Induced pluripotent stem cells → Used to research diseases

The method of introducing the reprogramming factors into the specialised adult cells requires careful consideration. Certain types of viruses can be used as vectors to introduce the factors but the genes may not function in the same way as embryonic stem cells and can result in the formation of tumours. Other ways to introduce genes may be safer but less successful. The potential advantages are that if in the future safety can be guaranteed the stem cells can be used as a bank to replace lost, damaged or diseased tissues in many parts of the body. Until then they are only used in research to model diseases of certain types of cells.

Adult Stem Cells

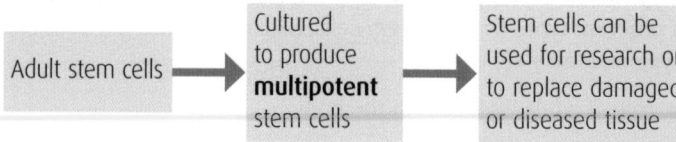

Adult stem cells → Cultured to produce **multipotent** stem cells → Stem cells can be used for research or to replace damaged or diseased tissue

> **DON'T FORGET**
>
> Eggs contain mitochondria which have their own DNA.

> **DON'T FORGET**
>
> Pluripotent cells can differentiate into all types of specialised cell. Multipotent can differentiate into one or a few types of specialised cell.

THERAPEUTIC VALUE OF STEM CELLS

Therapeutic value looks at the potential of stem cells in medicine. Stem cells are of special interest in the repair of diseased or damaged organs and to replace lost tissue.

Skin grafts

If a person is badly burnt and has lost most of their skin there is not enough good skin to use in grafts. A solution is to remove some adult stem cells from an area of good skin. The stem cells can then be cultured in the laboratory to produce skin cells. The new skin can then be grafted onto affected areas on the patient. The skin will not be rejected but isn't perfect as it lacks the complexity of normal skin not having hair follicles and sweat glands.

Bone marrow transplantation

Bone marrow stem cells are multipotent and can produce blood cells and platelets. Transplantation is used to treat certain types of blood related cancers, leukaemia or sickle cell anaemia. The patient's own bone marrow stem cells are destroyed and then replaced with healthy bone marrow stem cells either from a compatible donor or from the patient themselves if their bone marrow was harvested when they were healthy.

Cornea repair

Damaged corneas are more usually removed and replaced with healthy corneas from dead donors. This procedure is tried and tested but is invasive and suffers from a global lack of donors. Scientists have found that **multipotent** stem cells are located at the edge of the cornea. They can produce corneal or conjunctival cells. Stem cells can be removed from the patient's 'good' eye and be transplanted onto the damaged eye. Studies have included the use of contact lenses as culture media for the stem cells.

ETHICAL ISSUES OF STEM CELL USE

Certain control measures are in place to try to address some of the ethical implications of stem cell use.

Moral

Unused blastocysts from embryonic stem cell lines are destroyed as they are not allowed to develop beyond day 14 of the development stage. This is when the embryo would normally implant in the uterus and develop into a foetus.

Health

A complete medical history of adult stem cell donors is required to minimise the chance of recipients developing other medical problems.

Safety

Stem cells must be safe to use in the treatment of patients, i.e. they should not cause other conditions or diseases such as tumours caused by certain types of iPSCs. It is for this reason that ongoing research and thorough testing is vital.

 ONLINE

Find out more about the uses of stem cells at www.brightredbooks.net

 ONLINE TEST

Head to www.brightredbooks.net and test yourself on this topic.

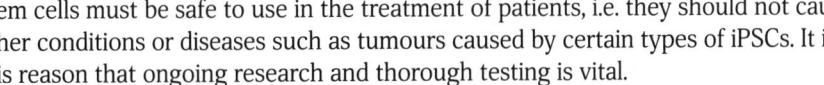 **THINGS TO DO AND THINK ABOUT**

1   Why must embryonic stem cells be used before they reach day 14 of the development stage?

2   What is the problem with using induced pluripotent stem cells for therapeutic use?

MUTATIONS

Any alteration to the genetic code is called a mutation. These are the results of **random** and **spontaneous** changes.

REGULATORY SEQUENCE MUTATIONS

Some regions on chromosomes are regulator genes. They may code for small **repressor** proteins that block expression of other genes on the chromosome. Their advantage is that they stop large proteins from being produced when they are not needed. In this way the organisms save energy and resources. Other regulator genes code for **activator** proteins that promote the transcription of some genes. Mutation in regulator genes can, therefore, change the expression of genes, resulting in absent or excess proteins and changing the phenotype (characteristics) of the organism; such mutations may even be lethal.

SPLICE SITE MUTATIONS

A mutation at a splice site may mean that some introns are not cut out but remain in the mRNA sequence during post-transcriptional modification. These introns will be read during translation and the protein will have extra amino acids which will alter the sequence and, therefore, function of the protein.

SINGLE GENE MUTATION

Individual genes are affected by point mutations. A gene is a segment of a DNA molecule that carries the code for a protein. The code takes the form of a sequence of bases, with triplets of bases coding for amino acids. If the sequence of bases is altered, the corresponding sequence of amino acids may change, possibly altering the protein produced. Types of gene mutation:

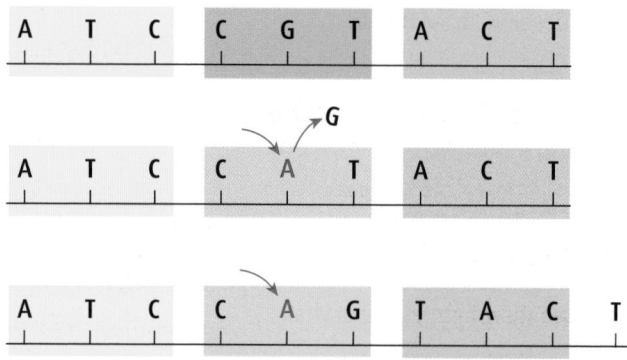

Normal strand of DNA

Substitution – one of the bases is replaced by a different base (here A replaces G). This can result in **splice site, nonsense** and **missense** mutations.

Insertion – an extra base is inserted into the sequence, in this example A, moving the bases after the insertion one place to the right. This is a **frame-shift mutation**, as every amino acid after the point of insertion is altered.

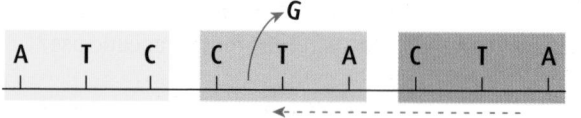

Deletion – a base is removed from the sequence, in this example G, shifting the bases one place to the left. This is another **frame-shift mutation**, as every amino acid after the point of deletion is altered.

CONSEQUENCES OF POINT MUTATIONS

DNA triplets code for specific amino acids. There are 20 amino acids but 64 triplets (from the possible combinations of the four nucleotide bases A, T, C and G arranged in sets of three). So, an amino acid may be coded by several triplets. This means that a mutation can have any one of five consequences:

1 Silent: a nucleotide is substituted, but the new triplet codes for the same amino acid and the protein is normal.
2 Neutral: the substituted nucleotide results in a triplet that codes for a different amino acid. However, it is similar to the original and the mutated protein still functions normally.
3 Missense: the substituted nucleotide results in a triplet that codes for a different amino acid, which changes the function of the protein.
4 Nonsense: the substituted nucleotide results in a stop codon, so the polypeptide chain is shorter.
5 Frame-shift: this is the result of a deletion or insertion mutation and the entire sequence of triplets after this point will be different, meaning that every amino acid after the mutation will be affected.

DON'T FORGET

A substitution can result in a code for a different amino acid, but sometimes the new triplet of bases codes for the same amino acid and so the mutation has no effect.

IMPORTANCE OF POINT MUTATION IN EVOLUTION

Point mutations are a source of **variation**. Occasionally, the new form (allele) of the gene can result in a protein that gives the individual a survival advantage. If they have a better chance of surviving to reproduce, the new alleles will increase in frequency in new generations. Over time, populations show change as the frequency of new phenotypes produced by mutations overtakes the frequency of less desirable alleles – evolution is happening.

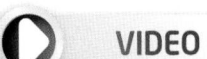

VIDEO LINK

Watch the first 6 minutes and 24 seconds of the YouTube clip on mutations at www.brightredbooks.net

CHROMOSOME STRUCTURE MUTATIONS

Whole sections of chromosomes can be altered, which affects several genes.

DUPLICATION	DELETION
Genes from one of a pair of homologous chromosomes transfer to the other, leading to duplication of genes. Myoglobin and haemoglobin are thought to have evolved from a common ancestral gene that duplicated and subsequently mutated.	Sections of DNA are left out during replication which means that genes are lost from the chromosome.

DUPLICATION
Gene 1 Gene 2 Gene 3 Gene 4 / Gene 7 Gene 8
Gene 5 Gene 6
Gene 1 Gene 2 Gene 3 Gene 4 Gene 5 Gene 6 Gene 5 Gene 6 Gene 7 Gene 8

DELETION
Gene 1 Gene 2 Gene 3 Gene 4 Gene 5 Gene 6 Gene 7 Gene 8
Gene 3 Gene 4
Gene 1 Gene 2 Gene 5 Gene 6 Gene 7 Gene 8

INVERSION
A section of DNA breaks from the DNA and is rotated **180 degrees** before reattaching. This results in the sequence of some genes being reversed.

Gene 1 Gene 2 Gene 3 Gene 4 Gene 5 Gene 6 Gene 7 Gene 8
Gene 1 Gene 2 Gene 3 Gene 4 Gene 5 Gene 6 Gene 7 Gene 8
Gene 1 Gene 2 Gene 3 Gene 5 Gene 4 Gene 6 Gene 7 Gene 8

TRANSLOCATION
Sections are swapped between different chromosomes

Gene 1 Gene 2 Gene 3 Gene 4 Gene 5 Gene 6 Gene 7 Gene 8
Gene A Gene B Gene C Gene D Gene E Gene F
Gene A Gene B Gene C Gene D Gene E Gene F Gene 6 Gene 7 Gene 8

THINGS TO DO AND THINK ABOUT

1 Redraw these chromosomes to illustrate the following chromosome structure mutations:
   a translocation
   b duplication
   c deletion
   d inversion

2 A section of DNA has the following sequence: ATGCAGTAC.
   What type of gene mutation is represented by each of the following altered sequences?
   a ACGCAGTAC
   b ATGCATAC
   c ATGCATGTAC

ONLINE TEST

How well have you learned this topic? Take the test at www.brightredbooks.net

CHANGES TO CHROMOSOME NUMBER

The chromosome number may be changed. One chromosome may be absent from or additional to the gamete haploid number, or there may be entire extra sets of chromosomes.

DON'T FORGET

Haploid (n) = single set of chromosomes. All gametes are haploid.
Diploid (2n) = two sets of chromosomes. Human somatic cells are diploid.

POLYPLOIDY

Sometimes complete sets of chromosomes fail to separate during cell division and one daughter cell gains an extra set of chromosomes – cells are normally diploid (2n) but polyploid cells have complete extra sets of chromosomes, so are 3n or above. This condition happens when there is a complete breakdown of the cell spindle. Polyploidy is extremely rare in animals but is more frequent in plants.

POLYPLOIDY IN PLANTS

Polyploidy in plants can be extremely useful both to the plant and, commercially, to humans. It can be induced experimentally by heat or cold shock, or by exposure to some chemicals, such as **colchicines**. Plants with cells containing three or more complete sets of chromosomes often grow larger than those plants with the normal diploid complement of chromosomes. This is of economic importance in cultivated plants, such as wheat and coffee, whose polyploid varieties produce a much higher yield than the diploid variety.

When polyploidy is produced by crossing two different varieties of plant, the resulting hybrid often shows improved yield, drought resistance, growth or resistance to disease. These improvements in hybrids are collectively called **hybrid vigour**.

The diagram below shows how polyploidy can occur in plants.

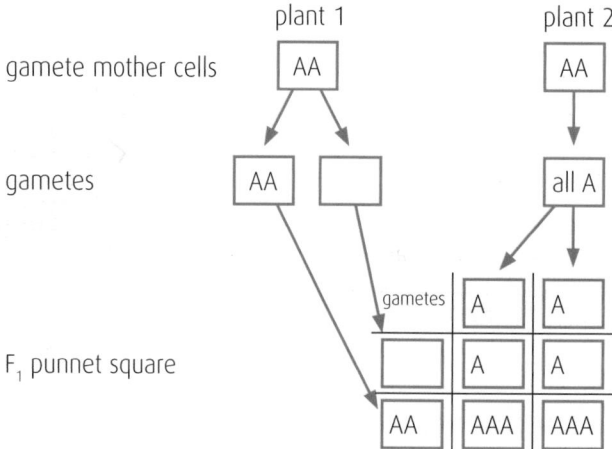

EVOLUTIONARY SIGNIFICANCE OF POLYPLOIDY IN THE PRODUCTION OF FOOD CROPS

The natural occurrence of polyploidy in certain crop plants has led to the evolution of many modern plants that are important to our economy. Examples include banana, potato, swede, oil seed rape, wheat and strawberry

Polyploid variants of the wild mustard plant were collected by early farmers who, through artificial selection of desirable characteristics, were able to develop various green vegetables that are in common use today.

broccoli

cauliflower

brussel sprouts

kohlrabi

wild mustard

cabbage

kale

POLYPLOIDY EVOLUTION IN ANIMALS

Polyploids are sometimes found among certain species of fish and amphibians. Polyploidy in animals is useful to humans as it can be induced artificially, for example to produce sterile rainbow trout to stock fishing areas. Polyploidy in mammals is often lethal. A rare example of a polyploidy mammal is the plains viscacha rat, which is a tetraploid.

Although it is rare today, polyploidy is thought to have played a major role in the evolution of vertebrates 500 million years ago.

ONLINE

Try some polyploidy quizzes and view an animation at www.brightredbooks.net

 THINGS TO DO AND THINK ABOUT

1 Explain why polyploidy in plants may be of use to humans.

2 Complete this cross

diploid plant   tetraploid plant

gamete mother cells   2n   4n

gametes

ONLINE TEST

How well have you learned this topic? Take the test at www.brightredbooks.net

EVOLUTION

Evolution is the gradual change in populations which results from variation and selection of certain alleles over generations.

VERTICAL INHERITANCE: SEXUAL REPRODUCTION IN EUKARYOTES

Vertical inheritance describes the passing of genomic information from parents to offspring. The process is slow as it involves the stages of gamete production and fertilisation, and may be influenced by seasons and mating rituals. It results in variation by mixing up parental genotypes. Gametes show variation because of the random way that homologous pairs of chromosomes separate to form haploid gametes.

Sexual reproduction

P phenotype	Red Petals	Red Petals		R – red petals
				r – white petals
P genotype	Rr	Rr		
gametes	R r	R r		

Punnett square:

	R	r
R	RR	Rr
r	Rr	rr

F_1 punnet square

F_1 genotypic ratio — 1 RR : 2Rr : 1 rr

F_1 phenotypic ratio — 3 red petals : 1 white petals

VERTICAL INHERITANCE: ASEXUAL REPRODUCTION

This type of inheritance is much faster than sexual reproduction as it does not involve the production of gametes or fertilisation. However, it does not produce the variation achieved by the meeting of gametes from two parental genomes.

Parent

Parent genotype — Aa

Clone genotype (offspring) — Aa

HORIZONTAL TRANSFER

Prokaryote to prokaryote

Prokaryotes can transfer genetic material horizontally to other prokaryotes and this is a means to promote variation. Bacteria can transfer plasmids to other bacterial species. This means that bacterial genomes can be shared rapidly and they can undergo rapid evolution. Consider that transgenic bacteria reproduce asexually and can double in number every 20 minutes!

contd

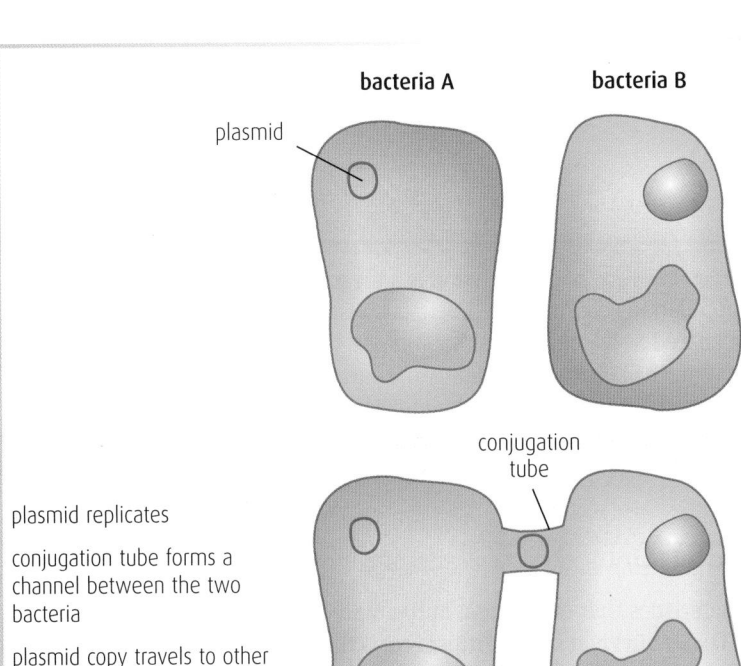

bacteria A   bacteria B

plasmid

conjugation
tube

plasmid replicates

conjugation tube forms a
channel between the two
bacteria

plasmid copy travels to other
bacterium

conjugation tube
disintegrates

transgenic bacterium
contains new genetic
information

DON'T FORGET

Horizontal transfer is the
mechanism that introduces
variation into prokaryotes.

VIDEO LINK

The clip at
www.brightredbooks.net
describes the mechanisms
involved in introducing
variation in more detail.

VIRUS TO EUKARYOTE

Some viruses can insert their genome into the host cell's genome. Many non-coding
introns are thought to be the result of viral genetic integration.

Example:

The HIV retrovirus can be
spliced into the DNA of
the host white blood cell.

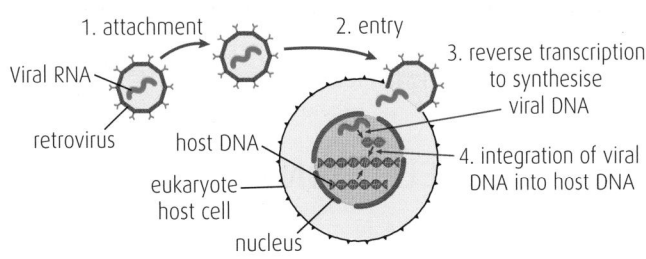

1. attachment   2. entry

Viral RNA

retrovirus

host DNA

eukaryote
host cell

nucleus

3. reverse transcription
to synthesise
viral DNA

4. integration of viral
DNA into host DNA

VIDEO LINK

Find out how the HIV
retrovirus inserts its DNA
into white blood cell DNA at
www.brightredbooks.net

THINGS TO DO AND THINK ABOUT

1 Draw a mindmap to summarise the mechanisms involved in producing variation in:

   a eukaryotes

   b prokaryotes

ONLINE TEST

Once you've learned
about this topic, test your
knowledge at
www.brightredbooks.net

SELECTION

MUTATION

Random mutation occurs in every population and is a means of introducing **variation** into a species. When subpopulations cannot interbreed, because an isolation barrier exists, new genes cannot be shared between groups. Sometimes a mutation occurs that benefits an individual, making it more likely that the individual survives, breeds successfully and produces more offspring. The mutation gives the individual a **reproductive advantage** and the frequency of the mutated allele will increase in successive generations.

NATURAL SELECTION

Individuals with characteristics that make them better suited to their environment are more likely to breed. This is **survival of the fittest** and is the basis for **natural selection**. The term 'fittest' might refer to a fast prey organism that can outrun a predator, or a plant that has become drought resistant in a dry habitat.

Natural selection depends on a population that cannot be completely sustained by the environment and a selection pressure which means that individuals with beneficial alleles are more likely to survive and reproduce than those with weaker characteristics. This process is **non-random** (not due to chance) and, over generations, causes a decrease in the frequency of undesirable alleles and an increase in desirable alleles.

DON'T FORGET

Natural selection can affect gene frequencies in two ways:
Advantageous mutations = non-random increases in alleles.
Deleterious mutations = non-random reductions in alleles.

LOSS OF HARMFUL MUTATIONS IN POPULATIONS

A **deleterious** (damaging) DNA sequence will eventually disappear from a population. Some mutations are lethal and the early death of the affected individual means that the sequence disappears rapidly. This type of *negative* selection is also **non-random**.

SEXUAL SELECTION

Female animals produce very few eggs when compared to males, who produce millions of sperm during their fertile life. The female needs to ensure that her offspring have the best chance of survival, given the time and energy she invests in their development and growth. She will mate with the strongest male and may base her selection on his ability to out-compete other males or protect their territory. Alternatively, her selection may be based on a male's skills in nest building, performing mating rituals, or on having elaborate decorations such as long tail feathers.

VIDEO LINK

The videos at www.brightredbooks.net illustrate some of the behaviours or phenotypes employed by male birds to get a mate.

STABILISING SELECTION

This type of selection gives an advantage to individuals that possess an average phenotype.

Human birth weight

Babies with a very low birth weight are more likely to develop life-threatening conditions. Babies with a very high birth weight are more likely to experience complications during birth, endangering both mother and infant. Babies within the average size range are more

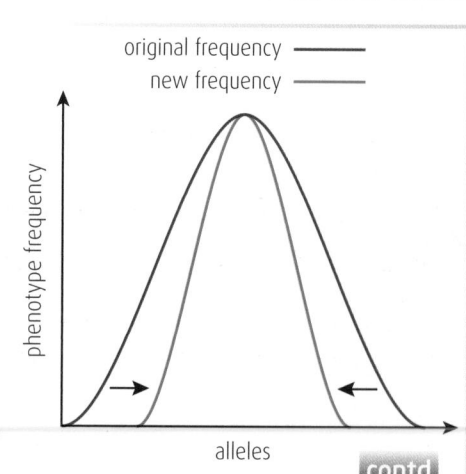

original frequency ——
new frequency ——

phenotype frequency

alleles

contd

likely to be healthy and to have an uneventful birth. Average size babies, therefore, have a better chance of surviving to reproductive age and passing on their alleles.

DISRUPTIVE SELECTION

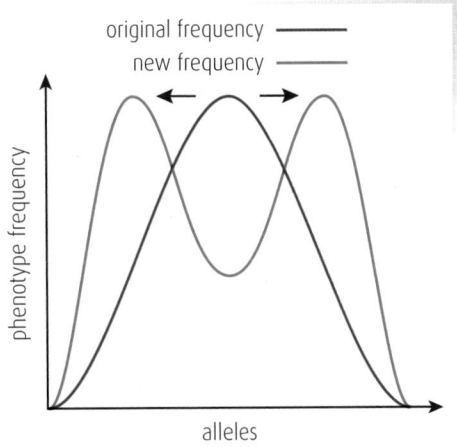

This type of selection favours **two** extreme forms of a characteristic.

Daphnia

Daphnia are tiny freshwater organisms. A population in a lake was found to be infected with a parasite. A subsequent change in phenotype frequencies was observed, with two distinct populations emerging; each displaying different characteristics. One population produced individuals that could withstand infection but had poorer reproductive success: its low birth rate was balanced by a low death rate. The other population succumbed more easily to infection, but was able to reproduce faster. Daphnia which were in the middle range decreased in number as they succumbed to the disease and had no reproductive strategy to compensate for their death rate.

DIRECTIONAL SELECTION

This type of selection favours alleles at one phenotypic extreme and occurs in response to **changing** environmental pressures. This causes a progressive change towards a characteristic with successive generations.

Industrial melanism of the peppered moth

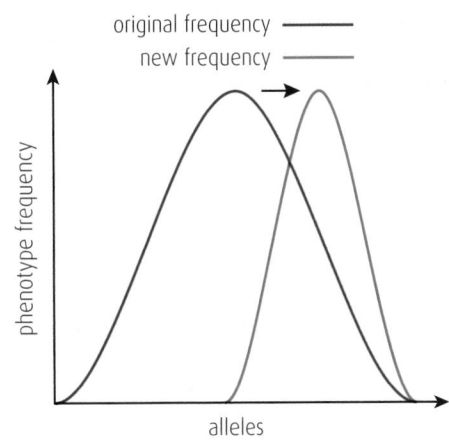

The industrial revolution in Britain caused a rapid increase in air pollution. Trees got darker in colour as tree lichens died and black soot was deposited. The peppered moth was commonly light in colour but, in industrial areas, became more visible and was more frequently eaten by birds. The rarer, dark (melanic) variety was better camouflaged, giving it a better chance of surviving to reproduce. The dark allele increased in frequency over successive generations. So, directional selection favoured a move from the light colour to the dark colour.

 THINGS TO DO AND THINK ABOUT

1   What kind of selection is shown in the evolution of a long neck in giraffes and what selective advantage did it confer?

2   Rewrite the following statement and delete the incorrect option.

Natural selection involves random/non-random changes in gene frequencies. The changes are the result of selective/non-selective pressures. Stabilising selection favours an extreme/average phenotype. Disruptive selection favours one/two phenotype extreme(s). Directional selection favours one/two phenotype extreme(s).

 ONLINE TEST

Head online and test yourself on this at www.brightredbooks.net

GENETIC DRIFT

All the gene sequences (alleles) in a population form the **gene pool**. The frequency of these sequences remains fairly constant in a large population, but **random** effects cause fluctuations from one generation to the next. Over time, there may be a slow drift with one allele increasing at the expense of another. Such drift is much more obvious in small populations and may result in the increase or loss of some alleles due to **the founder effect** and **neutral mutations**. These changes can be observed as differences in phenotype (physical characteristics such as feather colour, height of stem, molecular structure of protein).

DON'T FORGET

An allele is the form of one gene, for example, the gene for fur length has the alleles for short or long fur.

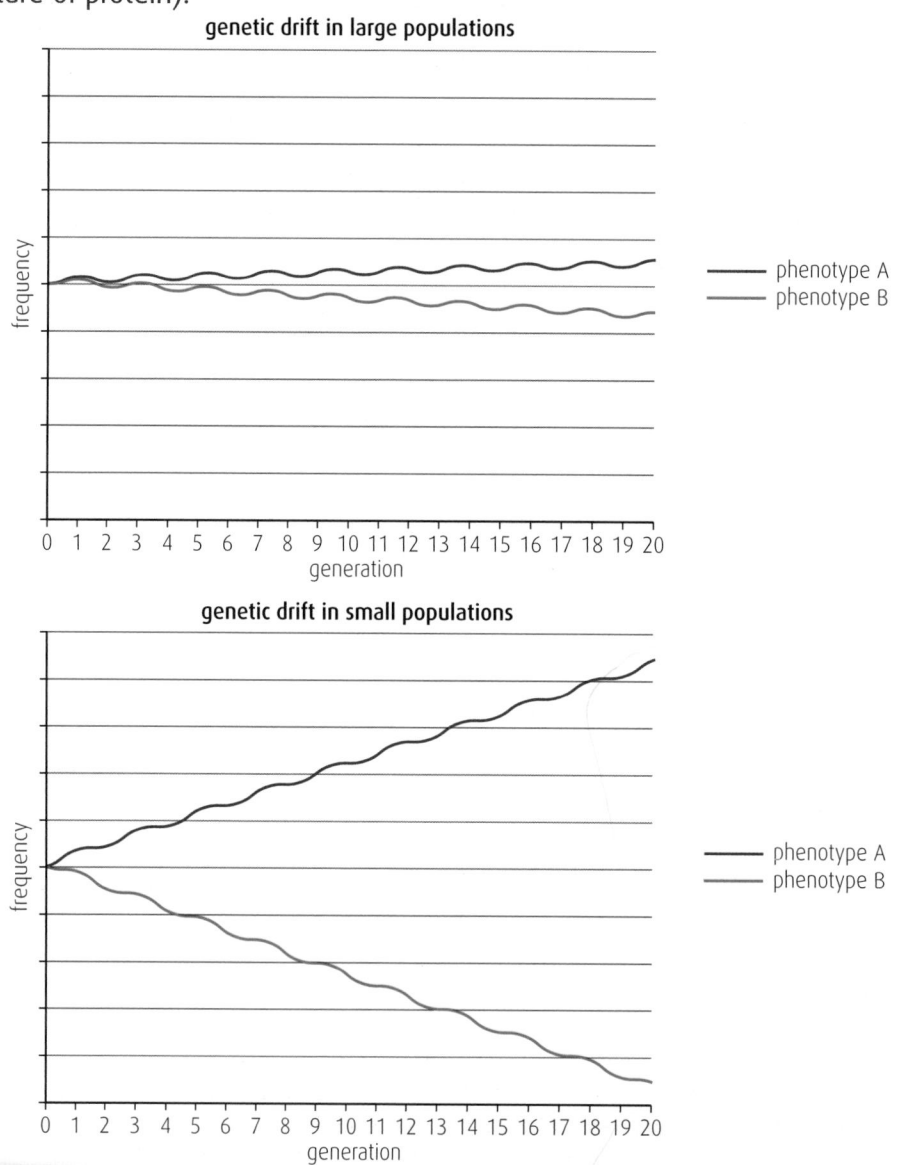

THE FOUNDER EFFECT

If a small **splinter** group becomes isolated from the main population, the individuals may take with them a disproportionate number of certain alleles, perhaps missing others out completely. Thus, the allele frequency is changed – recessive alleles may represent a higher than normal percentage of the total. The result is that the breakaway group **founds** a new population with different allele frequencies to the original; the subsequent population will reflect the gene pool of the splinter group but not that of the original, larger population. This is known as the **founder effect**.

GENETIC DRIFT AND NEUTRAL MUTATIONS

Some mutations have a **neutral** effect on an individual, for example a substitution mutation may result in a different, but very similar, amino acid in a sequence. However, the combined effect of more than one neutral mutation could be quite different, resulting in advantages or disadvantages, i.e. you may get away with one altered, but similar, amino acid in a sequence but two are more likely to alter the protein's properties. These mutations are rare in large populations and, so, it is even rarer for two individuals with neutral mutations to meet and breed.

However, if a splinter group breaks away from the main population, these neutral mutations may represent a higher percentage of the alleles in the new, smaller founding population. This increases the chances of two individuals with neutral mutations meeting and producing offspring who possess both mutations. Selective pressures may result in the increased frequency of these alleles until they eventually become the norm.

original
population

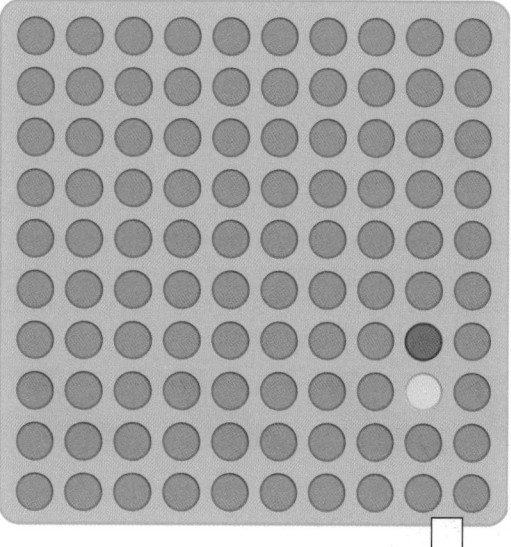

normal alleles are red

blue and yellow represent two different neutral mutations

population = 100

blue mutation =1

yellow mutation =1

normal = 98

Probability of blue meeting yellow = 1 in 99

splinter group

founder
population

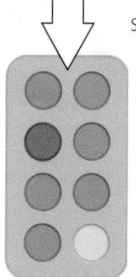

population = 8

blue mutation =1

yellow mutation =1

normal = 6

probability of blue meeting yellow = 1 in 7

THINGS TO DO AND THINK ABOUT

The Founder Effect is illustrated by the inhabitants of a tiny isolated island in the Pacific Ocean. In 1775 a typhoon struck the island, wiping out about 90% of the population and leaving about 20 survivors. One of the survivors carried the allele for complete colour blindness; a condition that leaves the individual unable to see any colour. Given the tiny size of the surviving population and the fact that they were cut-off from other people it was inevitable that, after a couple of generations, couples would meet who were cousins. Thus parents who were descended from the original carrier had children with the condition. Today 1 in 12 inhabitants has the condition. An affected person can see only black, white and shades of grey. This has certain everyday consequences such as being unable to see green mould on food. More seriously, the person is also dazzled by bright daylight and has some degree of visual impairment.

Questions on genetic drift

1   How are changes in allele frequency due to genetic drift different from those due to natural selection?

2   Describe the impact of the founder effect on genetic drift.

 ONLINE

Read up on the founder effect and genetic drift at www.brightredbooks.net

 ONLINE TEST

How well have you learned this topic? Take the test at www.brightredbooks.net

SPECIATION

If populations are isolated, the gene flow between populations is halted. Spontaneous, **random** mutations (which are different in each population) mean that such populations become phenotypically different. Selection occurs and after many generations, the changes in alleles and possible changes to chromosome structure may mean that the populations can no longer interbreed to produce fertile offspring. They have become different species. The formation of new species by the process of evolution is called **speciation**.

DON'T FORGET

Speciation depends on three factors occurring in sequence:
1 isolation
2 mutation
3 natural selection

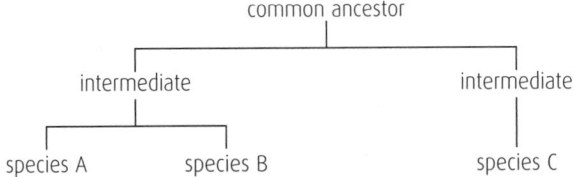

ALLOPATRIC SPECIATION

This occurs when groups are separated from the main population by a **geographical barrier**, such as a river, desert, mountain or sea. Over time, the groups evolve through genetic mutation and natural selection into populations that are genetically different. This difference means that, if members of the two groups meet, they can no longer interbreed to produce fertile young. They are two distinct species.

The barrier is removed and the two populations can mix.

The two populations cannot interbreed. They have become separate species

barrier

random mutation and natural selection

Time

SYMPATRIC SPECIATION

Sometimes new species form even when there is **no physical barrier** separating groups in a population.

- **Ecological barrier**: if habitats start to change, the preferred habitat of one group may isolate individuals from other groups. Changes can include differences in moisture level, temperature and pH.

- **Behavioural barrier**: reproductive differences between the subpopulations can prevent interbreeding. For example, subpopulations may flower at different times or there may be a lack of attraction between males and females of different subgroups.

Example:

In plants, when homologous pairs of chromatids fail to separate during gamete formation and polyploid offspring are produced (containing extra sets of chromosomes). The polyploids cannot interbreed with the original plants as their gametes are incompatible.

Example:

In animals, if a new resource is introduced to an area that causes individuals to separate from each other.

contd

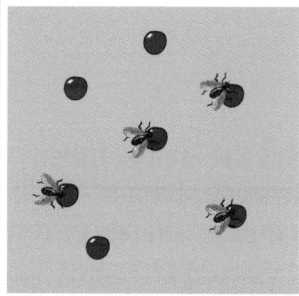

The common ancestor of the apple maggot fly laid its eggs in the fruit of the hawthorn.

Apple trees were introduced to the area and some flies laid their eggs in the apples.

Flies tend to breed and lay their eggs on the same type of fruit that they came from. Thus, there were two groups of fly that **avoided** each other and, over time, evolved into two separate species; the hawthorn maggot fly and the apple maggot fly.

HYBRID ZONES

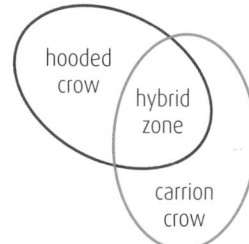

A hybrid zone is an area where two populations of closely related species meet and interbreed to produce fertile, hybrid offspring. For example, the hooded crow is mostly found in north-west Scotland and the carrion crow predominates in the rest of Scotland. They were thought to be the same species but recently have been reclassified as being closely related but different species who can breed together to produce fertile hybrids. In areas where they are found together, they will recognise each other as potential mates. The hybrids are less vigorous than their parents, giving them a selective disadvantage that means they are less likely to be successful and spread out of the hybrid zones. The incidence of hybridisation is decreasing, leading to speculation that the populations are undergoing speciation and separating.

VIDEO LINK

Watch the video at www.brightredbooks.net for a description of speciation and to find out more about hybrid zones.

 ## THINGS TO DO AND THINK ABOUT

ONLINE TEST

How well have you learned this topic? Take the test at www.brightredbooks.net

A new underground species of mosquito has been identified in the London Underground. This species shares many common alleles with the above-ground species and is thought to have branched off from it. The underground environment provided an artificial climate that was very different from above-ground conditions and was the isolating mechanism between the two populations.

The above-ground species is inactive in cold seasons, is very sociable, mates in swarms in spring/summer and feeds on birds. The underground species is active and breeds all year round in the warm, humid underground environment, mates in private and feeds on humans and rodents.

Questions on speciation

1 What are the three stages of speciation?

2 Explain the role of isolating mechanisms in speciation.

3 Give an account of the isolating mechanisms involved in:
   a allopatric speciation
   b sympatric speciation

GENOMIC SEQUENCING

All of the genes in an organism are known collectively as the **genome**, and the study of the genome is known as **genomics**. Technology has enabled scientists to determine the nucleotide sequence of specific genes and the entire genome of certain species. Nucleotide sequences can be compared in a branch of science called **bioinformatics** which employs detailed computer and statistical analyses. A comparison of sequence data has revealed that different species are surprisingly similar, which suggests that the genome is **highly conserved**.

Species 1	Species 2	Genomic similarity
Human	Chimp	96%
Human	Mouse	85%

Species of importance to humans have been targeted first for sequencing: crop plants, farm animals, crop pests and organisms that cause disease in humans or food species. The information may be used, for example, to develop genetically enhanced crops or to find alternatives to the chemical control of agricultural pests and pathogens.

VIDEO LINK

Look at a study of human evolution and the Neanderthals at www.brightredbooks.net

MOLECULAR CLOCKS AND PHYLOGENETICS

Phylogenetics compares **sequence data** of different species to determine how **related** they are and when they diverged (split) from a common ancestor. The length of the branch is proportional to the number of mutations in the sequence.

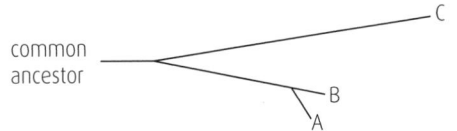

Species C has more differences in sequence, so is most distantly related to the common ancestor.

Species A and B have more similarities with each other, so diverged from each other later.

Here is a phylogenetic tree showing primate evolution:

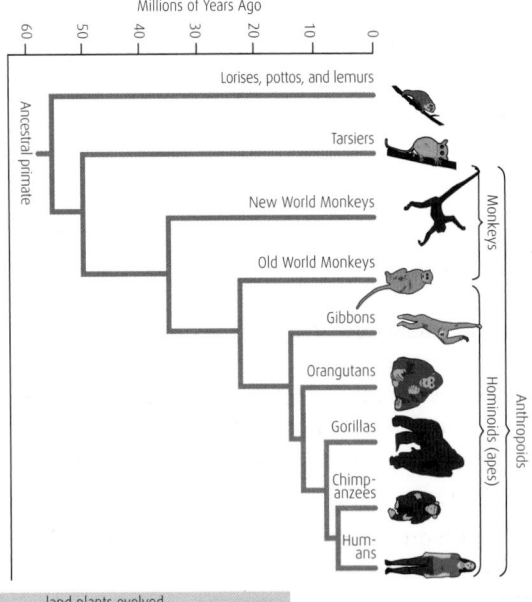

The **molecular clock** is based on the principle that, when groups separate, they become genetically different because of random mutations. So, the higher the number of mutations observed, the longer the time since they diverged. Research has identified three distinct domains of living things that split from the first **universal ancestor:**

1 bacteria

2 archaea (prokaryotes that survive in extreme conditions, such as deep sea thermal vents)

3 eukaryotes (plants, animals and fungi).

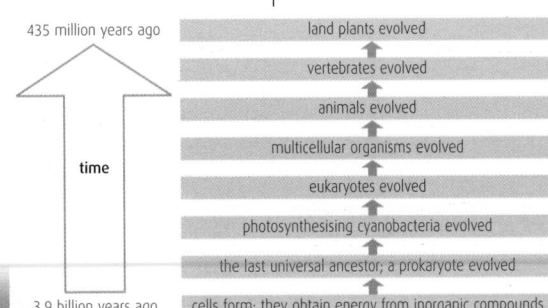

MAIN SEQUENCE OF EVENTS IN THE EVOLUTION OF LIFE

A timeline in the evolution of life has been proposed, using fossil dating evidence combined with genomic sequencing.

PERSONAL GENOMICS

The human genome sequence was completed in 2003 and provides a reference database. Individuals all have slight variations in their sequences, which can provide information about their risk of developing certain diseases or conditions. Assessing risk is a complex task as mutations are not always **harmful** but may be **neutral**. Variations in individuals could influence the effectiveness of certain medicines or increase the risk of developing serious side effects of treatment.

An individual's personal genome sequence could, therefore, be used to predict the risk of developing conditions, allowing them to make lifestyle choices to reduce the risks; it could also be used to inform the prescription of the most effective drugs. This type of personalised medicine is known as **pharmacogenetics**.

It is important to note that the development of a disease is often due to a *combination* of genetics and modifiable factors, such as diet, activity and stress levels. Genetic information could, therefore, be misinterpreted if taken out of context. There are also ethical implications regarding the right of access to genetic information: prospective employers? insurers?

THINGS TO DO AND THINK ABOUT

Copy the timeline below onto graph paper.

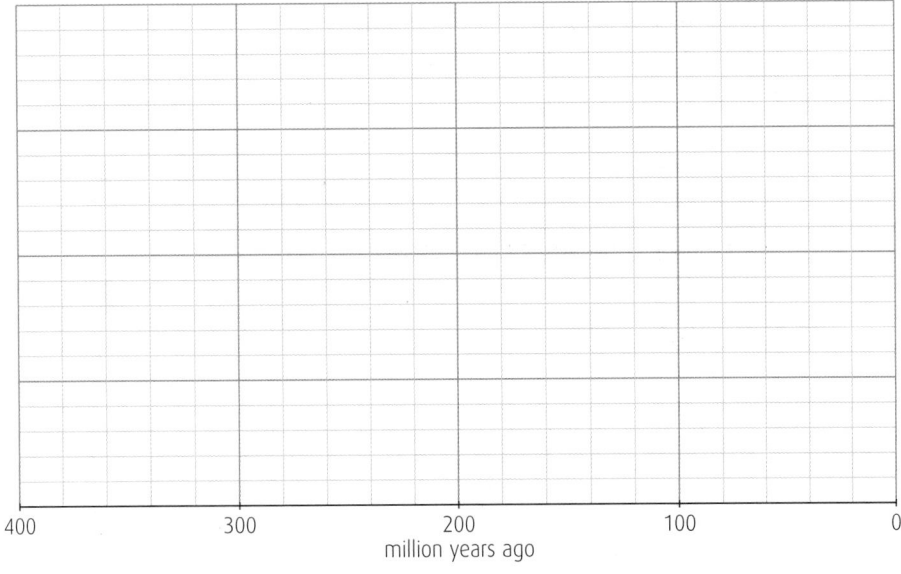

400   300   200   100   0
million years ago

1 a   Construct a phylogenetic tree from the following information.
       Fern evolution
       There is evidence of common ancestor ferns dating back to 380 million years ago (m.y.a.). Two branches diverged 360 m.y.a. Whisk ferns and ophioglossoid ferns diverged from one branch 320 m.y.a. Marattioid ferns, horsetails and lepto ferns diverged from the second branch 350 m.y.a.

   b   Which of the following fern species has more mutations when compared to its common ancestor: whisk ferns or horsetails?

   c   What is the name of the technique that estimates the time of evolutionary divergence based on the extent of mutation?

2 Place these evolutionary events into the correct order: vertebrates evolve, photosynthesising cyanobacteria evolve, land plants evolve.

3 Should a person's admittance to study medicine (a stressful occupation) be based on their genomic sequence in addition to other entry criteria? Does the faculty have a duty of care to the prospective student?

 **ONLINE TEST**

Head online and test yourself on this at www.brightredbooks.net

METABOLISM AND SURVIVAL

MEMBRANES

Both prokaryote and eukaryote cells are surrounded by a cell membrane. Eukaryotes also have membrane-bound structures (organelles) lying within the cell. These are either enclosed within a single membrane (for example, endoplasmic reticulum, Golgi apparatus, vesicles) or are surrounded by two membranes (chloroplasts, nucleus and mitochondria).

Each organelle is a small compartment which enables the cell to:

- separate digestive enzymes from the rest of the cell, preventing damage to the cell
- localise the metabolic activity of the cell by storing metabolites used in a particular cell reaction close together
- maintain high concentrations of metabolites; the convoluted shapes of organelles provide a large surface area to volume ratios within relatively small compartments.
- maintain high reaction rates, due to the relatively large surface area over which reactions can occur.

MEMBRANE STRUCTURE

The membrane is made up mainly of a double layer (a **bilayer**) of **phospholipid** molecules. Scattered on and within the membrane (and forming a mosaic pattern) are **proteins** of varying size. The position of the molecules within the membrane is not fixed and they constantly drift, giving the membrane a fluid-like nature. Hence, this model is referred to as the **fluid mosaic model**.

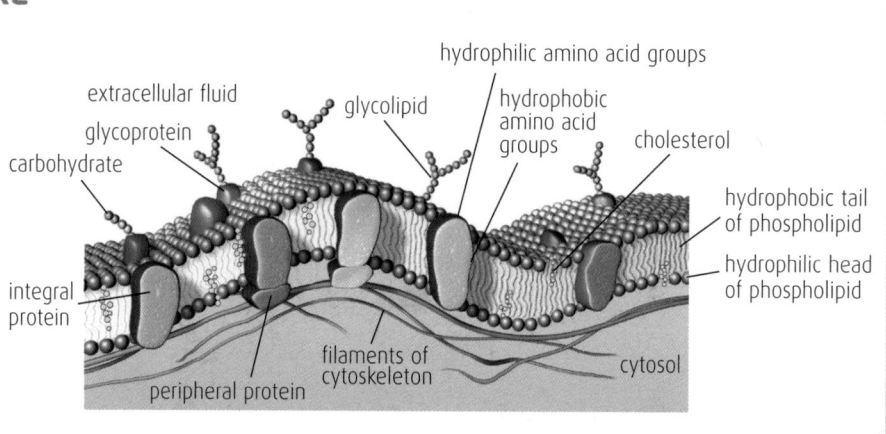

ROLES OF MEMBRANE PROTEINS

Membrane proteins have important functions, including acting as transport molecules and as enzymes.

Transport

Some molecules, such as oxygen and carbon dioxide, can move across the membrane from a high to a low concentration by passing through the phospholipid bilayer. This form of transport is called **diffusion**. It is passive and does not require energy.

Other ions and molecules that cannot pass between the phospholipid molecules require the presence of membrane transport proteins. These take the form of protein pores and carrier proteins (protein pumps). You should be able to discuss the following roles of membrane proteins.

membrane enzymes	Some membrane proteins act as enzymes to catalyse reactions.
protein pores	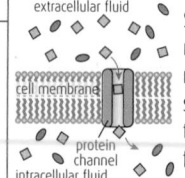 Some proteins act as pores through which ions and molecules can move passively from one side of the membrane to the other. These channel-forming proteins span the membrane and have a central water-filled pore through which substances pass. Each channel protein facilitates the transport of one specific ion or molecule.

> **DON'T FORGET** ➕
>
> As temperature increases, molecules move faster and the rate of diffusion increases.

contd

protein pumps

Active transport requires ions and molecules to be pumped against a concentration gradient. Energy from the breakdown of ATP is used to change the shape of the protein pump, allowing the molecule (or ion) to move across the membrane.

One example of a protein pump is the sodium–potassium pump. The carrier protein actively pumps sodium ions out of the cell and also pumps potassium ions into the cell. This pump creates an electrical gradient (for every three Na⁺ ions that leave the cell, only two K⁺ ions enter, making the cytoplasm relatively negatively charged). This is important to muscle and nerve cells in the production of action potentials, and in the production of a chemical gradient which assists the transport of essential nutrients, such as glucose, into the cell. The mechanism of the sodium–potassium pump is shown below.

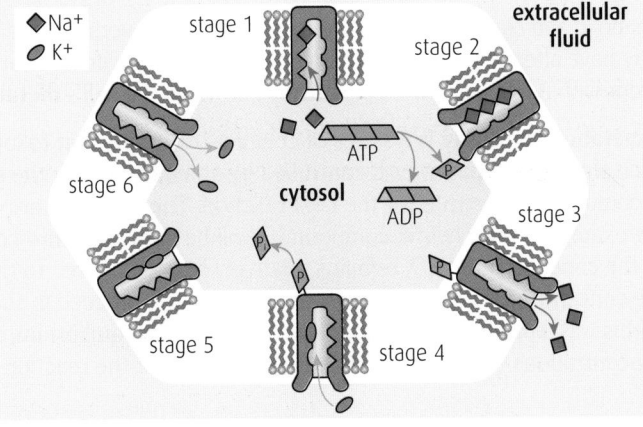

Stage 1: three Na⁺ ions bind to the carrier protein.

Stage 2: after conversion of ATP to ADP, a phosphate binds to the carrier protein causing a conformational change.

Stage 3: the Na⁺ ions are released on the extracellular side of the membrane.

Stage 4: two K⁺ ions move in and bind to the carrier protein, causing the phosphate to be released from the carrier protein.

Stage 5: when the phosphate has been released, the carrier protein returns to its original conformation.

Stage 6: the K⁺ ions are released on the intracellular side of the membrane.

THINGS TO DO AND THINK ABOUT

Study the two graphs showing the effect of temperature and oxygen concentration on active transport of sodium ions.

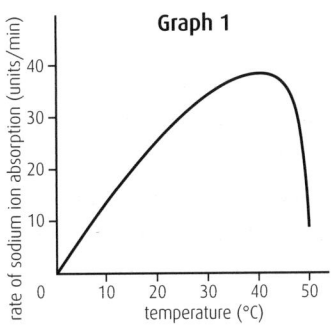

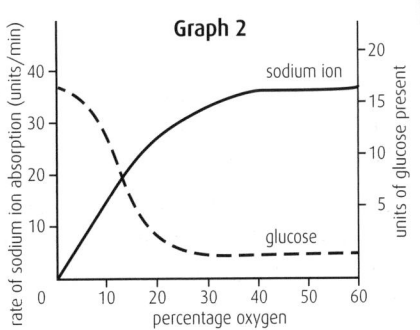

ONLINE TEST

Once you've learned about this topic, test your knowledge at www.brightredbooks.net

Use the graphs to answer the following questions.

1   Describe how temperature and oxygen concentration affect the rate of sodium ion transport across the membrane.

2   Explain why transport of sodium ions is associated with a decrease in glucose concentration.

ENZYMES 1

METABOLIC PATHWAYS

Within the cell, reactions can be classed as:

- **anabolic reactions** – large molecules are synthesised from several smaller ones, with energy being used up

- **catabolic reactions** – large molecules are broken down into several smaller molecules, usually with a release of energy.

Anabolic and cathabolic pathways can have both reversible and irreversible steps and may have alternative routes. The sum of all the anabolic and catabolic reactions that occur within a living cell is collectively known as the cell's **metabolism**.

A **metabolic pathway** is a series of chemical reactions that follow on, one after another. Each stage in the pathway is controlled by an **enzyme**, with the product of one reaction becoming the substrate for the next reaction. The reactions are often reversible. In the example shown below, compound X will be converted into compound Y, as long as the concentration of X remains relatively high compared to that of compound Y. If the concentration of X decreases, the reaction will proceed in the opposite direction. In this way, enzymes act to drive reactions towards equilibrium, where the relative concentrations of molecules at the start and end of the reaction are balanced.

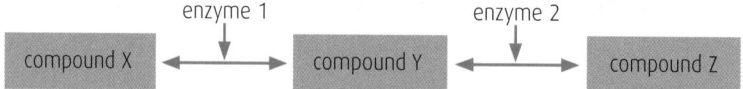

Where several enzymes are involved in catalysing different stages of a metabolic pathway, the enzymes may be associated together in a cell membrane, producing a **multi-enzyme complex**. By locating close together within the membrane, the overall rate of reaction of the whole pathway can be increased, making the pathway more efficient. **DNA polymerase** and **RNA polymerase** both form parts of multi-enzyme complexes.

ACTIVATION ENERGY

At normal body temperatures, chemical reactions would take place at too slow a rate to maintain life if enzymes were not present.

Chemical reactions involve the breaking of bonds and the formation of new ones. To start a reaction, energy (the **activation energy**) must be used to break bonds within the reactant molecules. As energy is absorbed, bonds become increasingly unstable. At their most unstable, the molecules are said to be in a **transition state**. When the bonds in the reactants break, the molecular structure of the product forms. Enzymes act by reducing the activation energy required to reach the transition state and, therefore, allow reactions to take place at lower temperatures.

ENZYME ACTION

Each enzyme can only act on one substrate. Enzyme action is, therefore, said to be **specific**. This is because the shape of the substrate molecule fits into the enzyme's active site. When two or more substrate molecules are involved in the reaction, the molecules fit into the active site in the particular **orientation** required to allow the molecules to react.

INDUCED FIT

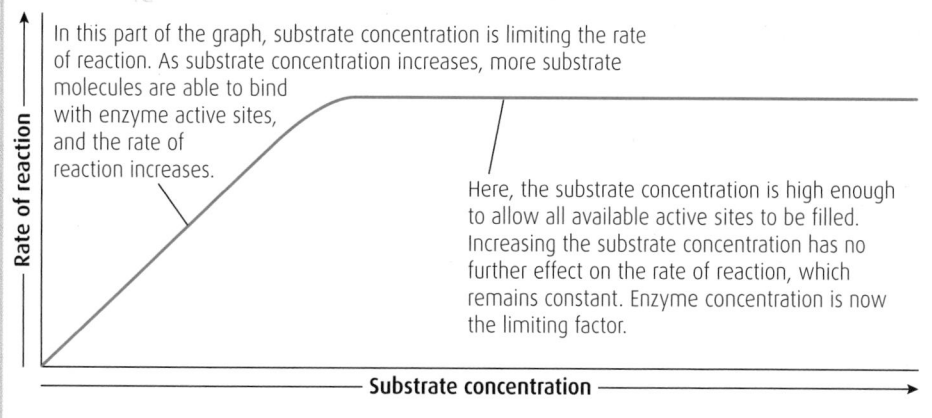

hexokinase enzyme

active site

glucose

the glucose induces a change in shape in the enzyme

enzyme–substrate complex

Enzyme function is described as an **induced fit**. The active site begins in an open position, which allows the substrate (with a **high affinity** for the active site) to move in and bind. Binding of the substrate causes a change in the shape of the active site to a closed position. This brings the substrate and enzyme closer together, increasing the chance of a reaction. Once the product has formed, the shape of the active site returns to the open position and the product (which has a **low affinity** for the active site) moves out.

SUBSTRATE CONCENTRATION

As the substrate concentration increases, the rate of reaction increases and then becomes constant. The graph below demonstrates how the rate of reaction is linked to the number of active sites that are filled at any one time.

In this part of the graph, substrate concentration is limiting the rate of reaction. As substrate concentration increases, more substrate molecules are able to bind with enzyme active sites, and the rate of reaction increases.

Here, the substrate concentration is high enough to allow all available active sites to be filled. Increasing the substrate concentration has no further effect on the rate of reaction, which remains constant. Enzyme concentration is now the limiting factor.

Rate of reaction

Substrate concentration

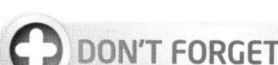 **DON'T FORGET**

To increase the rate of reaction further, the enzyme concentration would have to be increased.

 ONLINE TEST

Once you've learned about this topic, test your knowledge at www.brightredbooks.net

 THINGS TO DO AND THINK ABOUT

You should review the action of enzymes studied at National 5 level.

ENZYMES 2

COMPETITIVE INHIBITION

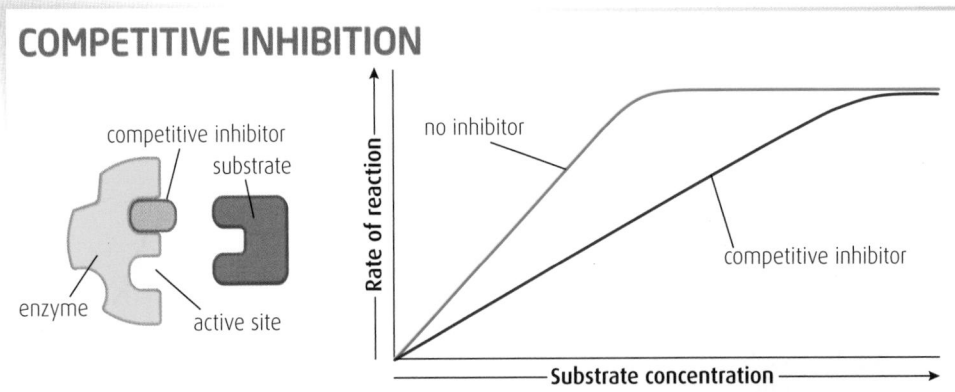

Competitive inhibitor molecules have a shape similar to that of the substrate. They bind with the enzyme's active site, preventing the substrate from entering. Because the substrate and inhibitor are in competition for the active site, increasing the substrate concentration causes an increase in the rate of reaction.

NON-COMPETITIVE INHIBITION

Non-competitive inhibitors bind to a part of the enzyme which is not the active site. As a result, the shape of the active site is altered and the substrate cannot enter. Because the substrate and inhibitor are not in competition for the active site, increasing substrate concentration has no effect on the rate of reaction. The rate of reaction remains low.

FEEDBACK INHIBITION (END PRODUCT INHIBITION)

The rate at which some metabolic pathways progress can be controlled by a build-up of the end product. In feedback inhibition, the end product binds to one enzyme at the beginning of the metabolic pathway, altering the shape of this enzyme's active site and stopping the pathway. This prevents too much end product from being produced. As the concentration of the end product drops, inhibition ceases and the pathway resumes again.

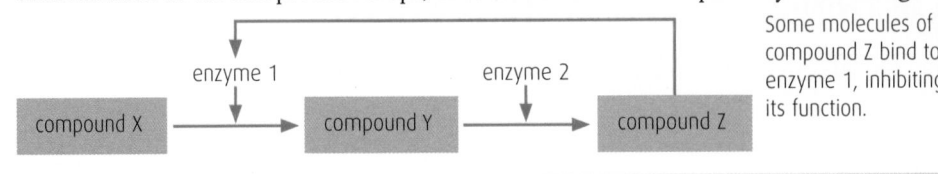

Some molecules of compound Z bind to enzyme 1, inhibiting its function

DEMONSTRATING FEEDBACK INHIBITION

You may have carried out the following experiment to study feedback inhibition using the enzyme phosphatase, which can be obtained by grinding down mung bean sprouts and centrifuging the mixture to extract a phosphatase enzyme solution.

VIDEO LINK

An animation showing end-point inhibition can be seen at www.brightredbooks.net

The enzyme is used in the following reaction:

$$\text{phenolphthalein phosphate} \xrightarrow{\text{phosphatase}} \text{phenolphthalein + phosphate}$$

To demonstrate that phosphate (the product of the reaction) can cause inhibition, test tubes are set up containing the substrate and different concentrations of sodium phosphate as shown opposite.

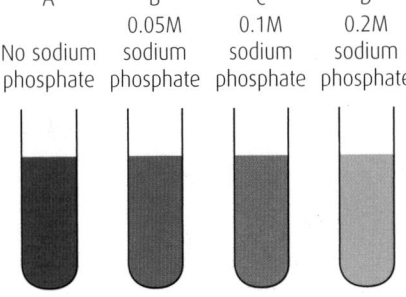

DON'T FORGET

Buffer is added to the test tubes to maintain a constant pH, increasing the validity of the experiment.

Each tube contains:

- phenolphthalein phosphate
- phosphatase (in the enzyme solution produced using the mung beans)
- buffer
- sodium phosphate (concentration shown above each tube).

The reaction is stopped by the addition of alkali (sodium carbonate). The alkali also causes the phenolpthalein to change from colourless to pink.

In tube A there is no inhibition and a high concentration of phenolphthalein is produced causing the solution to become a strong pink colour. As the concentration of phosphate increases in tubes B–D, inhibition of the enzyme phosphatase increases and less phenolphthalein is produced resulting in a less intense pink colour. It is possible to obtain quantitative results for this experiment by using a colorimeter to measure the transmission of light through the solution. The graph below show a typical set of results.

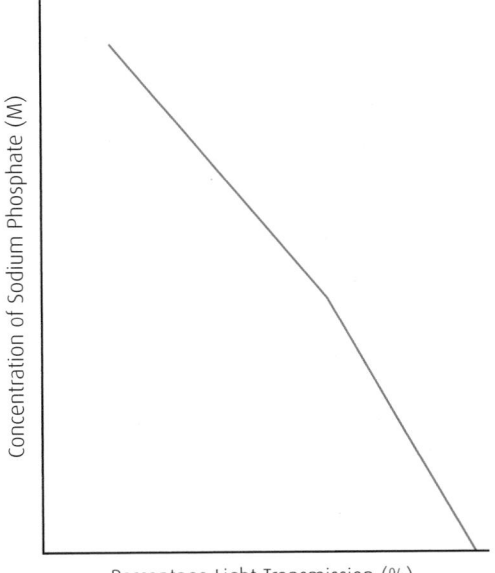

 ONLINE TEST

Head online and test yourself on this at www.brightredbooks.net

THINGS TO DO AND THINK ABOUT

1 Use the internet to research the action of the following poisons and toxins:

 a Poisons – hydrogen cyanide, lead, cyanide

 b Toxins – venoms, *Salmonella*

ENERGY RELEASE AND THE ROLE OF ATP

RESPIRATORY SUBSTRATES AND USES OF ENERGY

Molecules which can be broken down to release energy in respiration are called **respiratory substrates**. The energy released is used to fuel cellular processes such as protein synthesis, contraction of muscle, active transport, DNA replication, and carbon fixation.

Respiration occurs in cells of the three domains of life: Archaea, Bacteria and Eukaryota.

DON'T FORGET

During respiration, energy released by breaking bonds between carbon and hydrogen atoms is transferred in the form of electrons to electron carriers until the cell synthesises ATP.

ATP AND ADP

The series of reactions that make up respiration result in chemical energy being transferred to a molecule called **ATP**, adenosine triphosphate. ATP is a source of energy that can be used immediately by cells. During respiration, ATP is made when a bond forms between an **inorganic phosphate** (P_i) and **ADP**, adenosine diphosphate. This reaction is called **phosphorylation**. When the bond is subsequently broken, the energy is released and used in cellular processes, such as synthetic pathways.

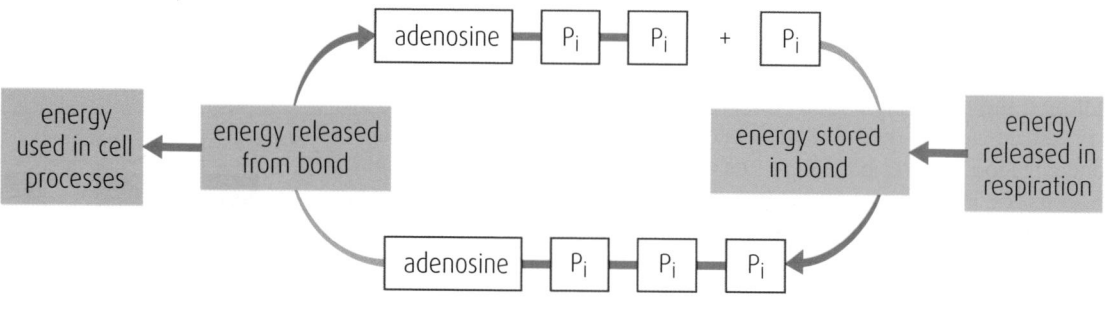

MITOCHONDRIA

Mitochondria are known as the power houses of the cell because they are the main site of ATP synthesis. A mitochondrion consists of two membranes: a smooth outer membrane and a folded inner membrane, surrounding a central **matrix**. The **citric acid cycle** takes place in the matrix. Most ATP is generated in the electron transport chain which consists of proteins embedded in the inner membrane. Folding of this inner membrane to form **cristae** increases the membrane's surface area, so that more molecules of the electron transport chain can fit in. The more folds there are, and the longer each fold is, the faster ATP can be produced. Very active cells contain lots of large mitochondria with many, long cristae.

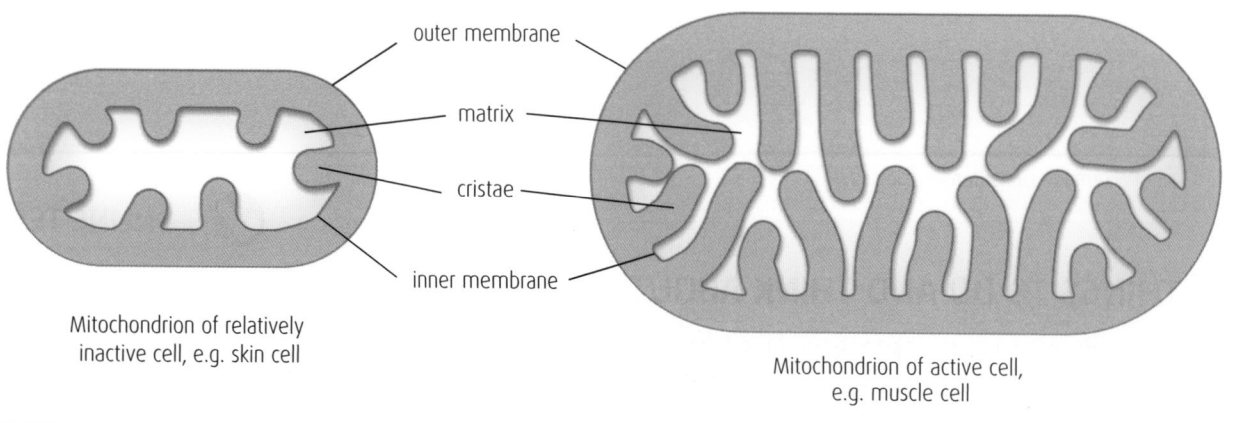

Mitochondrion of relatively inactive cell, e.g. skin cell

Mitochondrion of active cell, e.g. muscle cell

YEAST DEHYDROGENASE INVESTIGATION

Yeast cells contain the enzyme **dehydrogenase**, which removes hydrogen ions and electrons from molecules during respiration. Loss of hydrogen can be detected using **resazurin dye**, which changes colour from blue to pink and then to colourless as hydrogen is lost from the respiration pathway and added to the dye.

The table shows the results of an experiment where resazurin dye, yeast and glucose solution were mixed together and left to react in a warm water bath.

Test tube	Contents of test tube	Colour change	Explanation
A	resazurin dye boiled yeast suspension glucose solution	stays blue	Boiling the yeast suspension denatures the enzyme dehydrogenase and no reaction occurs.
B	resazurin dye live yeast suspension glucose solution	blue → colourless	Dehydrogenase in the yeast cells catalyses oxidation of glucose. The hydrogen released during respiration is added to the resazurin dye.
C	resazurin dye water live yeast suspension	blue → pink	Although no glucose is present to act as respiratory substrate, the yeast cells do contain a small quantity of food which acts as a respiratory substrate. Some oxidation will, therefore, occur and the dye becomes partially reduced.

Questions

Why should all solutions be pre-incubated in a warm water bath before being added together?

How could you modify the experiment to ensure that no stored carbohydrate remained in the yeast cells before using them?

THINGS TO DO AND THINK ABOUT

ATP is constantly being made in cells but the total mass in an organism is relatively constant. It is made when required and used almost immediately.

To show that ATP (and not glucose) can be used as a direct source of energy by the cell, glucose solution and ATP can be dripped onto teased out muscle fibres. You should be able to describe what would happen in each case.

Muscle cells do not contract when glucose solution is dropped onto them, but if ATP solution is used they will contract immediately.

ONLINE TEST

How well have you learned this topic? Take the test at www.brightredbooks.net

RESPIRATION 1

Respiration is a series of enzyme-controlled reactions.

DON'T FORGET

Dehydrogenase enzymes remove hydrogen ions and electrons and passes them to the coenzymes NAD and FAD.

VIDEO LINK

Watch the animation of fermentation at www.brightredbooks.net

DON'T FORGET

Fermentation in animals is a reversible reaction. Lactic acid is converted back to pyruvic acid when oxygen becomes available. This is called repaying the oxygen debt.

DON'T FORGET

Glycolysis results in a net gain of two ATP molecules.

STAGE ONE – GLYCOLYSIS

Glycolysis takes place in the cytoplasm of every living cell. No oxygen is required. It is divided into an **energy investment phase** where the conversion of 2ATP → 2ADP + 2Pi provides energy for the conversion of glucose to intermediate **phosphorylated** molecules (phosphate groups are added to the molecule). This is followed by an **energy pay-off phase** in which the intermediate molecules are converted into two **pyruvate** molecules. As four ATP molecules are produced during this phase, there is a net gain of two ATP (4ATP produced – 2ATP used) in glycolysis.

Hydrogen ions are released during the energy pay-off phase and are picked up by the hydrogen carrier molecule **NAD** to make **NADH**. The hydrogen ions will be passed on to the electron transport chain.

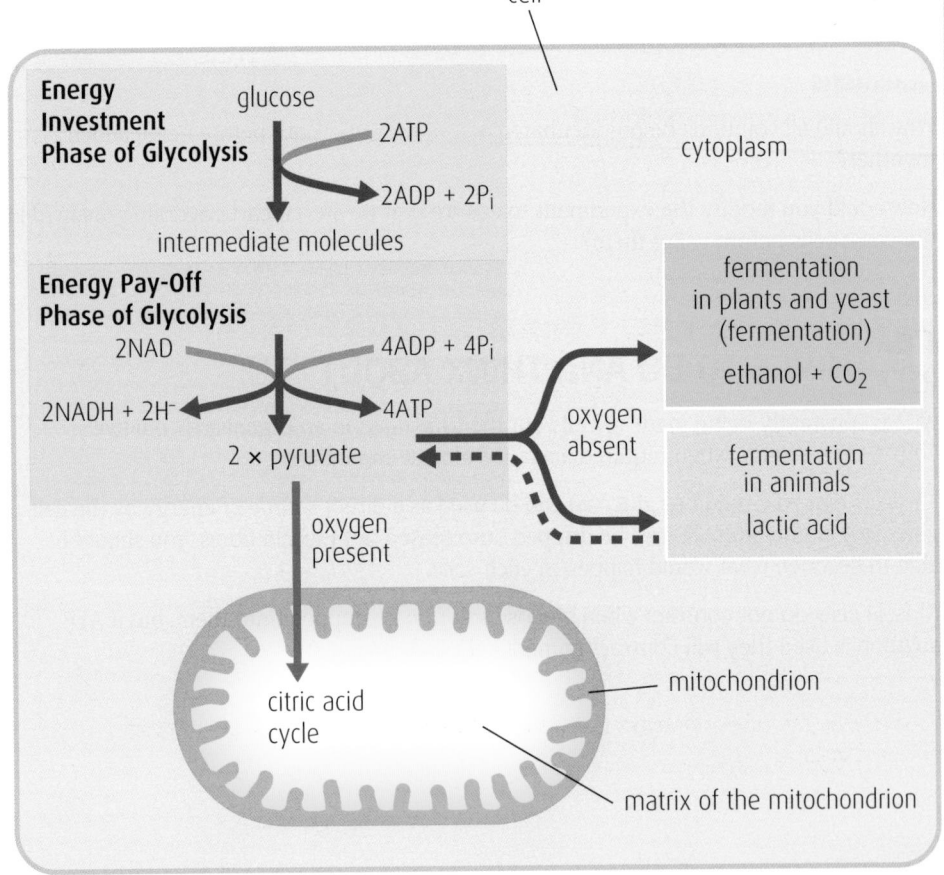

FERMENTATION

In the absence of oxygen, fermentation takes place. Fermentation produces only the **two ATP** molecules released in glycolysis.

In plants and yeast, pyruvic acid is converted to **ethanol and carbon dioxide**. In animals, **lactic acid** is produced.

STAGE TWO – CITRIC ACID CYCLE

The **citric acid cycle** takes place in the matrix of mitochondria. Breakdown of pyruvic acid produces carbon dioxide and an **acetyl group**. The acetyl group binds with **co-enzyme A** to produce **acetyl co-enzyme A**. The acetyl group and **oxaloacetate** then combine to produce **citrate**. As the cycle proceeds, carbon dioxide is released and dehydrogenase removes hydrogen ions and electrons which are picked up by either **NAD or FAD** to form **NADH** and **FADH$_2$** respectively. NADH and FADH$_2$ carry the hydrogen ions to the third stage of respiration, the electron transport chain, on the inner membrane of the mitochondrion.

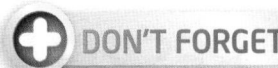
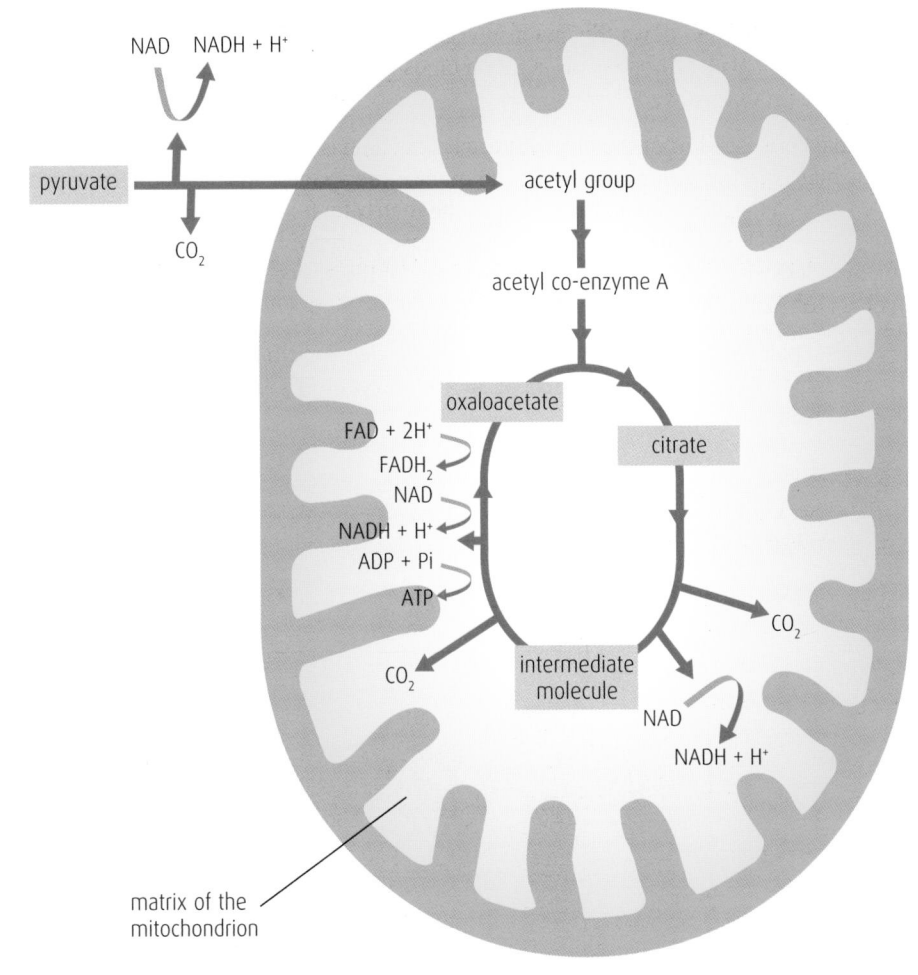

matrix of the mitochondrion

THINGS TO DO AND THINK ABOUT

1 Describe the role of dehydrogenase enzymes in both glycolysis and the citric acid cycle.

ONLINE

Work through the link on respiration at www.brightredbooks.net

ONLINE TEST

Once you've learned about this topic, test your knowledge at www.brightredbooks.net

RESPIRATION 2

ELECTRON TRANSPORT CHAIN

The electron transport chain is found on the **cristae** of the mitochondria and consists of a series of carrier proteins attached to the **inner mitochondrial membrane**. Hydrogen ions and electrons are transferred from NADH and $FADH_2$ to the electron transport chain. As the electrons are passed down the chain, energy is released. Electrons are finally passed to **oxygen** (the final electron acceptor) which binds with hydrogen ions and electrons in the matrix to produce **water**.

Energy that is released from the electron transport chain is used to pump hydrogen ions from the matrix to the **inner mitochondrial membrane space**, causing a concentration gradient to develop. Hydrogen ions return to the matrix by flowing through a channel in the enzyme **ATP synthase**. This flow of ions causes parts of the enzyme molecule to rotate in a clockwise direction, changing the shape of the active site and allowing the conversion of ADP + P_i to ATP.

Every time a pair of hydrogen ions pass through the electron transport chain, three ATP molecules will be produced. A total of **36 ATP** molecules is made in the electron transport chain.

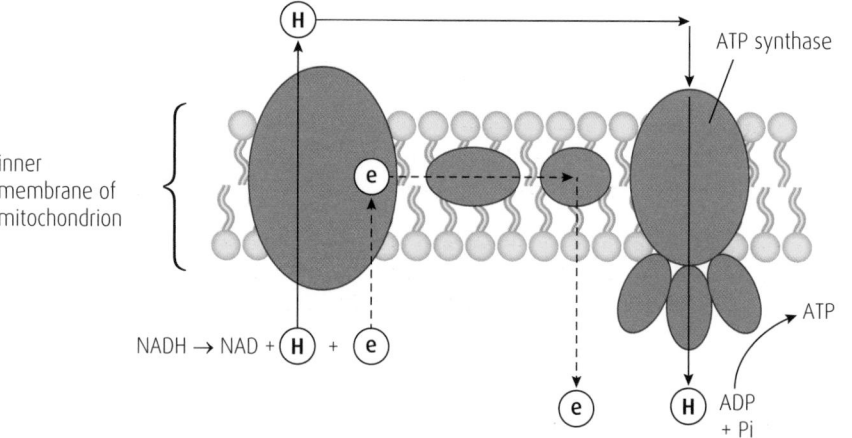

COMPARISON OF AEROBIC AND FERMENTATION RESPIRATION

	Aerobic respiration	Fermentation
oxygen required	yes	no
number of ATP molecules produced	38	2
waste products	carbon dioxide and water	in plants and yeast, ethanol and carbon dioxide
		in animals, lactic acid

ALTERNATIVE SUBSTRATES FOR RESPIRATION

Glucose is the main respiratory substrate. However, other carbohydrates, fats and proteins can be broken down and used as respiratory substrates when required as follows:

carbohydrates	glucose → used in glycolysis. starch and glycogen → broken down into glucose for use in respiration. sugars other than glucose → converted into either glucose or intermediate molecules for use in glycolysis.
fats	Fat contains twice the energy of either carbohydrates or proteins. It can be broken down into fatty acids and glycerol, both of which can enter the respiratory pathway. fatty acids → converted to acetyl co-enzyme A before entering the citric acid cycle. glycerol → converted to intermediate molecules for use in glycolysis.
proteins	While excess dietary protein can be used as an energy source, most of the protein that is taken in is used for growth and repair of body tissues. Proteins are broken down to produce amino acids. When excess amino acids are broken down, some of the products enter the respiration pathway. proteins → amino acids → converted to pyruvic acid, acetyl co-enzyme A, or intermediates in the citric acid cycle.

THINGS TO DO AND THINK ABOUT

1 Aerobic respiration is a more efficient pathway than fermentation as 38 ATP molecules are produced in aerobic respiration but only two ATP molecules are produced in fermentation. What happens to the rest of the energy in fermentation?

2 Alternative energy sources become important during periods of both marathon running and starvation. Use the internet to research the use of carbohydrates, fats, and proteins as respiratory substrates during marathon running and periods of starvation.

ONLINE TEST

Once you've learned about this topic, test your knowledge at www.brightredbooks.net

OXYGEN DELIVERY

Energy release in aerobic respiration requires oxygen to act as the final electron acceptor. It is, therefore, important that oxygen is able to rapidly reach all body cells. The development of a cardiovascular system in multicellular organisms allows this to occur.

CARDIOVASCULAR SYSTEM

You should be able to compare the cardiovascular systems of fish, amphibians, reptiles and mammals.

Fish

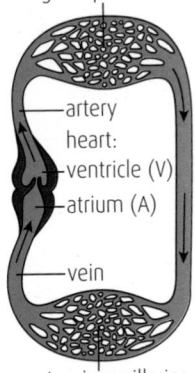

gill capillaries

artery
heart:
ventricle (V)
atrium (A)
vein

systemic capillaries

Circulation and blood pressure	Blood travels through a single circuit from the heart to the gills and then on around the body. Because the blood pressure decreases as the blood passes through the gills, body tissues receive blood under low pressure. Blood flow is aided, however, by body movements during swimming.	
Heart	Atrium	A single atrium receives deoxygenated blood from the body.
	Ventricle	A single ventricle pumps deoxygenated blood to the gills.

Amphibians and reptiles

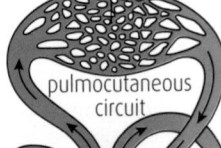
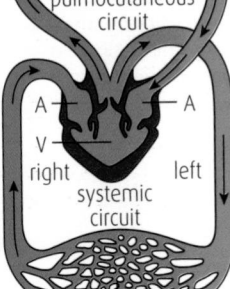

lung and skin capillaries

pulmocutaneous circuit

A   A
V
right   left
systemic circuit

systemic capillaries

Circulation and blood pressure	There is a double circuit with one circulation loop pumping to the lungs and skin, and another loop pumping to the body. Although blood loses pressure as it passes through the capillary beds in the lungs and skin, the pressure increases again as it passes through the heart for a second time.	
Heart	Atrium	A right atrium receives deoxygenated blood from the body.
		A left atrium receives oxygenated blood from the lungs and skin.
	Ventricle	A single ventricle receives blood from both atria. Mixing of oxygenated and deoxygenated blood can take place but is reduced by internal ridges that divert deoxygenated blood from the right atrium towards the lungs and skin and oxygenated blood from the left atrium towards the rest of the body.

Mammals and birds

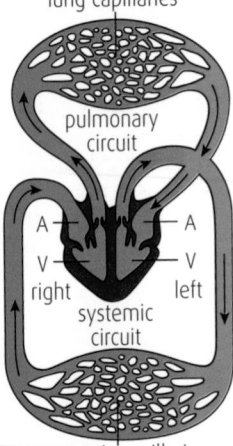

lung capillaries

pulmonary circuit

A   A
V   V
right   left
systemic circuit

systemic capillaries

Circulation and blood pressure	There is a double circuit with the right side of the heart pumping blood to the lungs and the left side pumping blood to the body. As with the amphibian and reptilian systems, pumping blood through the heart after it has been through the lungs ensures that the body tissues receive blood under high pressure.	
Heart	Atrium	A right atrium receives deoxygenated blood from the body.
		A left atrium receives oxygenated blood from the lungs.
	Ventricle	A right ventricle pumps deoxygenated blood from the right atrium to the lungs.
		A left ventricle pumps oxygenated blood to the body. By having two ventricles, mixing of oxygenated and deoxygenated blood is prevented. This is important to maintain the high metabolic rate required in these animals.

LUNGS

You should be able to compare the lungs of amphibians, mammals and birds.

Amphibians

Lungs in amphibians are small balloon-like sacs. This is an inefficient system as the surface area available for diffusion is low, being limited to the outer surface of the lungs. The amount of oxygen obtained through the amphibian lungs is small. To ensure that the animal has enough oxygen to support its metabolic rate, uptake is mostly by diffusion through the animal's moist skin and mouth.

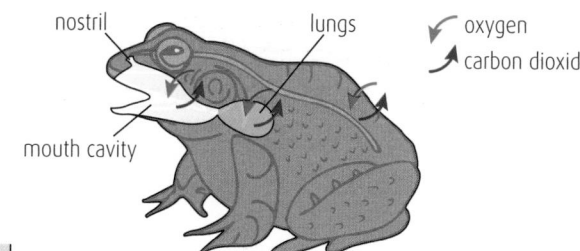

Reptiles

In reptiles, each lung is divided into several alveolar-type sacs, increasing the surface area over which diffusion can take place.

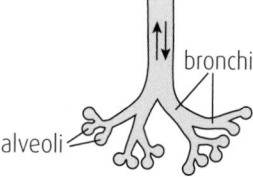

Mammals

The lungs contain millions of tiny alveoli which provide a very large surface area for diffusion. Breathing movements of the diaphragm and ribcage bring about emptying and filling of the lungs under negative pressure.

Birds

In birds, diffusion of oxygen takes place through a series of parallel tubes in the lungs. Air is passed through the tubes in one direction only (diagrams **a–d**), through the filling and emptying of air sacs which lie outside the lungs. Inhalation causes fresh air to be drawn into the posterior air sacs, while stale used air leaves the lungs to enter the anterior air sacs. On exhaling, the posterior air sacs push oxygen-rich air into the lungs and stale air from the anterior air sacs passes into the trachea.

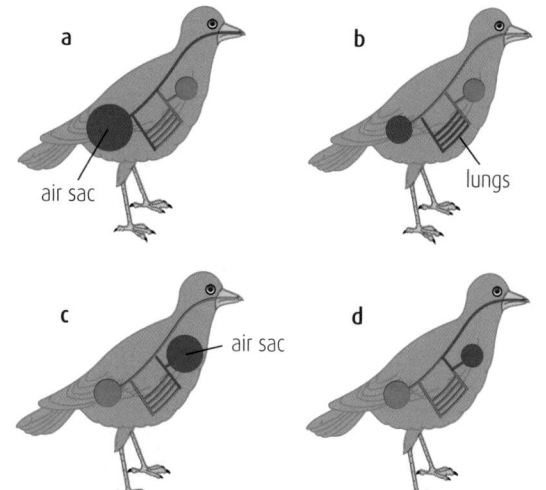

This unidirectional air flow prevents the mixing of inhaled and exhaled air, maximising the oxygen content of the air. This is important as flight at high altitudes has high metabolic needs in a low oxygen environment.

DON'T FORGET

Mammals and birds are said to have a *complete* circulation because there is no mixing of oxygenated and deoxygenated blood.

 THINGS TO DO AND THINK ABOUT

Maximum oxygen uptake – measuring fitness

VO_{2max} is the volume of oxygen used per kilogram of body mass per minute while exercising at maximum capacity (exercising flat-out). It is an indicator of fitness, VO_{2max} increasing with fitness level. To calculate VO_{2max} the ventilation rate, and oxygen and carbon dioxide concentration of inhaled and exhaled air are measured while the test subject exercises on either an exercise bike or a treadmill. The speed and intensity of the exercise are gradually increased, causing the oxygen consumption to increase. When oxygen consumption cannot increase further and becomes constant, VO_{2max} has been reached. Exercising to maximum capacity can be dangerous, particularly for non-athletes. A less accurate (but safer-to-measure) estimate of VO_{2max} can be determined using exercise tests that are slightly less strenuous.

 ONLINE TEST

Head online and test yourself on this at www.brightredbooks.net

ADAPTATION FOR LOW OXYGEN NICHE

DIVING ADAPTATION

Diving animals must cope with both a **limited supply of oxygen** and an **increase in pressure** with increasing water depth. As the depth of the dive increases, the increasing pressure causes an increase in the solubility of gases within the tissues, resulting in tissues becoming saturated. At high concentrations, both nitrogen and oxygen can have serious effects on the body. Nitrogen poisoning causes euphoria and delusions similar to a narcotic effect. The toxic effect of oxygen at high pressure can result in blackout or death.

As the diving animal surfaces, gases move out of the tissues and back into the blood. A human diver must ascend slowly enough to prevent these gases from coming out of solution and forming bubbles in the blood (the bends).

Adaptations in marine animals that allow them to dive successfully include changes to **blood flow**, **metabolism** and **an increase in oxygen storage capacity**. The table shows some of these adaptations.

Part of body	Adaptation	Benefit
lungs	Some deep-diving animals have compliant chest walls and lungs which collapse as the animal dives. Lung collapse begins in the alveoli, pushing air up into the larger air passages and preventing diffusion of gases into the blood. Specialised surfactant helps to re-inflate the lungs as the animal returns to the surface.	This keeps the gas tension relatively low, reducing the risk of oxygen toxicity, nitrogen narcosis and decompression sickness.
blood flow	Blood is diverted to the heart, brain and lungs. There is a decrease in both the cardiac output and total blood flow.	Blood is diverted to essential organs.
heart rate	Decreased	Assists in lowering metabolic rate.
red blood cells and haemoglobin	Red blood cells are stored in the spleen and released into the blood during deeper dives. This increases the haematocrit (percentage of red blood cells in a given volume of blood). The volume of blood is also relatively greater per unit of body mass in deep diving-animals. The increased number of red blood cells and larger volume of blood gives an increase in haemoglobin.	More haemoglobin is available for transport of oxygen
myoglobin	Diving animals have a myoglobin concentration in their muscles which is up to 30% greater than their land-living relatives.	Myoglobin provides a store of oxygen for active muscle cells.
metabolism	Some diving animals exhibit a lowered metabolism due to the need to respire anaerobically. This may be associated with increased concentrations of enzymes involved in the removal of lactic acid.	Accommodates the reduced synthesis of ATP in anaerobic respiration.

ADAPTATIONS TO HIGH ALTITUDE

At both sea level and at high altitude, the concentration of oxygen in the air is 21%. However, at high altitudes the atmosphere is less dense with gas molecules spaced further apart. This makes it more difficult for oxygen to diffuse through the alveolar wall into the blood, causing oxygen deprivation (hypoxia). Symptoms of high altitude sickness in humans include vomiting, headache, fatigue, and difficulty in thinking clearly. In severe cases, cerebral and pulmonary edema result from accumulation of fluid around the brain and haemorrhaging in the lungs. The increased stress placed on the lungs and cardiovascular system can also result in heart failure.

When humans live at lower altitudes and travel to a high altitude, the immediate response is an increase in the heart and breathing rates as the body works harder to get oxygen into the blood. Over the next few weeks, the body becomes acclimatised – by producing a greater number of red blood cells and developing a more extensive capillary network. Endurance athletes often train at high altitude in the run up to major championships in order to benefit from these physiological changes. However, within weeks of their return to sea level, the body returns to normal.

Humans who live constantly at high altitudes have evolved to survive. The table shows some of the adaptations that help with high altitude living.

Part of body	Adaptation	Benefit
lungs	Pulmonary capacity is increased. Breathing rate may be increased.	Allows for increased diffusion of gases.
haemoglobin	Some populations have increased levels of haemoglobin.	Oxygen transport can be maximised.
cardiovascular system	Red blood cell numbers are increased. There is an increase in the capillary network in muscles. Blood vessels are broader.	Oxygen transport can be maximised.

DON'T FORGET

Large multicellular organisms were able to evolve after atmospheric oxygen levels increased approximately 600 million years ago.

THINGS TO DO AND THINK ABOUT

Residents of the Tibetan plateau have normal levels of haemoglobin but have wider blood vessels and maintain a higher breathing rate. Genetic studies have shown that many Tibetans exhibit gene mutations close to the gene EPAS1, a gene which codes for a protein that is involved in sensing oxygen levels. It may be that the mutated genes are involved in responding to hypoxia.

Natives of Peru and Bolivia who live at high altitude show increases in both haemoglobin levels and lung expansion capability.

Look up the website www.xtreme-everest.co.uk to find out how scientists are studying human responses to high altitude in order to develop novel therapies.

ONLINE TEST

How well have you learned this topic? Take the test at www.brightredbooks.net

METABOLIC CONFORMERS AND REGULATORS

Metabolic processes proceed at different rates under different environmental conditions. For organisms to survive, abiotic factors such as temperature, salinity and pH must remain within a tolerable range. Animals can be divided into two groups: conformers and regulators.

DON'T FORGET ✚

Animals can be conformers for one factor and regulators for another.

CONFORMERS

Conformers are animals that allow their internal body conditions for a particular factor to vary with the external environment. They usually have low metabolic costs and narrow ecological niches. The animal must use **behavioural adaptations** to function effectively by either moving to advantageous conditions or by avoiding other less favourable conditions. Many conformers live in stable environments where there is little change to the abiotic factor.

For example, the body temperature of thermal conformers (i.e. animals that are ectothermic) is determined by the temperature of the environment in which they are living. This results in ectotherms being sluggish at low temperatures (remember that enzyme driven reactions progress slowly at low temperatures). Think of reptiles sitting on sun-warmed rocks to raise their body temperatures and therefore increase metabolic rate.

REGULATORS

Regulators are animals that use **physiological mechanisms** to maintain their internal body conditions at optimum levels (**homeostasis**). This requires a high energy input but allows the organism to function effectively in a much broader environmental range. For example, where an ectotherm must sit taking in heat from the surrounding rocks, a thermal regulator (endotherm) would already be functioning at an optimum internal temperature.

Homeostasis

Homeostasis is the maintenance of the body's internal environment in response to changes in the surroundings. This includes:

- water balance
- temperature
- blood sugar level

Negative Feedback

To allow homeostasis, there must be a corrective mechanism that acts when any variable in the internal environment changes too much. A mechanism like this uses negative feedback.

Aspects of the internal environment are monitored by receptor cells in monitoring centres around the body. Deviations from the normal level or set point (at which conditions are optimum for body processes) are detected by the receptor cells, and result in messages being sent from monitoring centres to effector organs. The effectors respond by bringing the level back to the set point. Messages sent out by the monitoring centres can be in the form of either:

DON'T FORGET ✚

Negative feedback works like the thermostat in a central heating system. When the temperature in your house goes too high (that is above the set point), the thermostat switches off the heating and then switches back on when the temperature goes too low (below the set point).

- hormones that are secreted into the blood, or
- nerve impulses.

The diagram summarises the events of negative feedback.

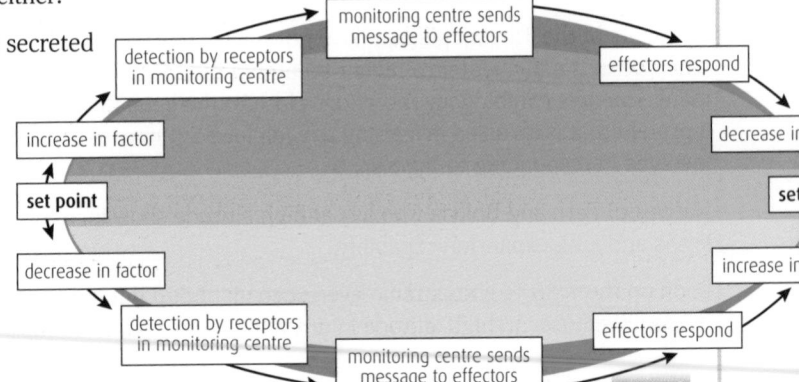

contd

Temperature Regulation

All chemical reactions in the body are controlled by enzymes. When body temperature is below the optimum temperature for enzyme function, metabolism is slow. Metabolism is fastest when the body temperature is close to the optimum temperature for enzymes. Above the optimum, enzymes start to denature; the metabolism slows down and eventually stops.

Receptors that detect blood temperature (thermoreceptors) are found in the lining of blood vessels in the **hypothalamus** – the temperature monitoring centre in the brain. When thermoreceptors detect changes in blood temperature, the hypothalamus sends out nerve impulses to effector organs in the skin and body muscles. The effectors bring about a response designed to return the temperature to normal. The diagram below summarises the process.

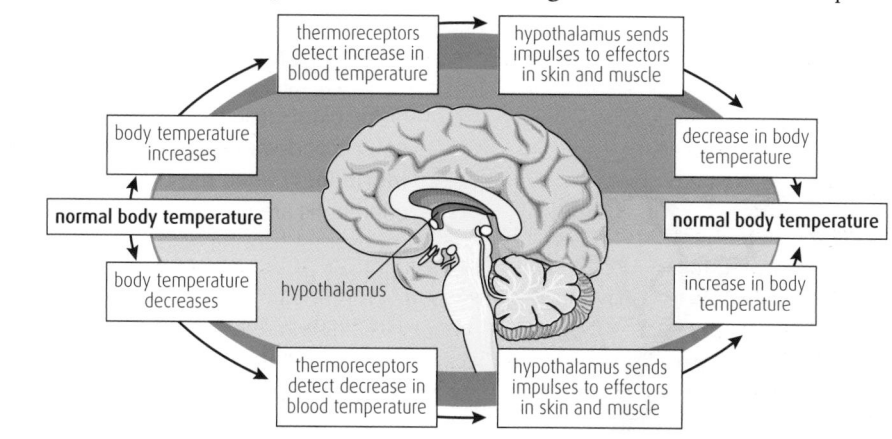

ONLINE

Look up the links at www.brightredbooks. net to view animations of body changes to temperature change.

DON'T FORGET

Body temperature can increase due to exercise, illness and exposure to a hot environment. Body temperature can decrease due to exposure to a cold external environment.

RESPONSE TO TEMPERATURE CHANGE

Responding to temperature change includes both voluntary and involuntary methods.

Voluntary responses

Voluntary responses include:
- removing clothes, opening windows, switching off heaters in response to increased body temperature, and
- putting on clothes, closing windows and switching on heaters in response to decreased body temperature.

Involuntary responses

Involuntary responses to temperature change are shown in the diagram on the right.

Response to decrease in body temperature		Response to increase in body temperature
vasoconstriction – constriction (narrowing) of skin arterioles diverts blood away from the skin's surface, reducing heat loss by radiation	skin arterioles	vasodilation – dilation (widening) of skin arterioles diverts more blood to capillaries at the skin's surface, increasing heat loss by radiation
decreased sweat production	sweating	increased sweat production causes water in the sweat to heat up on the skin's surface and evaporate, lowering body temperature
erector pili muscles contract, pulling body hairs into a raised position that traps air to insulate the body (like a duvet)	erector pili muscles	erector pili muscles are relaxed, allowing body hair to lie flat against the skin, minimising the insulating effect
shivering increases muscle activity and generates heat	shivering	no shivering
increases to generate heat	metabolic rate	decreases to reduce heat production

 ## THINGS TO DO AND THINK ABOUT

Where in the body is the temperature monitoring centre?

How does the temperature monitoring centre communicate with the effectors?

Why is body temperature important in carrying out metabolic processes?

Changes in surface temperature on the body can be measured using a thermistor. You may have carried out an investigation where thermistors were placed in the armpit (representing the body core) and between the fingers of one hand (representing the body shell), while the other hand was held in a bucket of icy water. Receptors in the skin detect a drop in temperature in the hand in the icy water and send nerve impulses to the temperature monitoring centre in the hypothalamus. The hypothalamus responds by stimulating vasoconstriction in both hands, diverting blood away from the body surface to conserve heat. A drop in temperature would be registered in the thermistor held between the fingers, as heat is conserved. The thermistor held in the armpit would register a constant temperature.

ONLINE TEST

Head to www.brightredbooks.net and test yourself on this topic.

SURVIVING ADVERSE CONDITIONS 1

Organisms can survive adverse conditions by:

- evolving mechanisms that allow them to survive in that environment
- entering a period of **dormancy** in which they minimise their metabolic rates to conserve energy at times of adverse environmental conditions.
- **migrating** to areas where conditions are more favourable.

DORMANCY IN PLANTS

Some plants lose their leaves and enter a period of rest and inactivity (**dormancy**) when the climatic conditions are not suitable for growth, for example during cold winters or summer drought.

The seeds of some plant species display periods of dormancy when they fail to germinate even under ideal conditions (in the presence of sufficient oxygen and water, and an appropriate temperature). This can be due to the presence of **chemical inhibitors** which prevent growth of the embryo, or to the presence of an **impermeable seed coat** which forms a physical barrier to the uptake of water.

DORMANCY IN ANIMALS

Hibernation

Hibernation allows animals to survive cold winter conditions and involves the lowering of **body temperature**, **heart** and **breathing rates**, and **oxygen consumption**. This lowers the **metabolic rate** at a time when food is limited and the animal slowly uses up its body fat reserves as the energy source. Homeostatic mechanisms remain active during hibernation, allowing the animal to respond to changes in the environment. For example, if the surrounding temperature drops suddenly, the animal's metabolic rate will rise to maintain its body temperature. This is important as it prevents the animal from freezing. Decreasing day length acts as the trigger for hibernation and the presence of a large enough fat reserve is also thought to be an important factor. The dormouse, bat and hedgehog are the only native British mammals that hibernate.

contd

Daily torpor

Some animals undergo short periods on a daily basis when their body temperature, heart and breathing rates, and oxygen consumption are lowered (**torpor**). This conserves energy between feeding bouts or may occur at night when there is a marked decrease in temperature. This type of behaviour is typically found in small animals with high metabolic rates, such as hummingbirds.

Aestivation

Aestivation is similar to hibernation but takes place in the summer or dry season during long periods of extremely high temperatures or extremely dry conditions, when food and water may be scarce. Terrestrial animals that aestivate often use underground burrows where the moisture level is higher and the temperatures are lower than on the surface. Aestivation requires less energy than hibernation to maintain a minimum body temperature and, so, the animal does not require a large store of body fat. Examples of animals that use aestivation are snails, earthworms, lizards and some tropical butterflies.

 **DON'T FORGET**

In predictive dormancy, the organism enters dormancy before the adverse conditions begin. In consequential dormancy, the organism enters dormancy after the adverse conditions begin.

THINGS TO DO AND THINK ABOUT

1   Suggest possible benefits of predictive and consequential dormancy?

2   Plant seed dormancy is an advantage to the plant as it prevents the seed from germinating when the seed is exposed to the wrong environmental conditions. However, it is possible to break dormancy through either scarification or pre-chilling. What does each process involve?

3   The African Bullfrog can be exposed to drought conditions at any time of the year. It survives by burrowing into the mud to form a chamber in which it remains dormant until the rains return.

   Which line in the table identifies the type and description of dormancy displayed here by the African Bullfrog?

 **ONLINE TEST**

Head online and test yourself on this at www.brightredbooks.net

	Type of dormancy	Description of dormancy
A	hibernation	consequential
B	hibernation	predictive
C	aestivation	consequential
D	aestivation	predictive

SURVIVING ADVERSE CONDITIONS 2

MIGRATION

In **migration**, animals avoid unfavourable conditions by moving from one place to another. This **regular pattern** of behaviour is displayed by birds such as the swallow and osprey, animals such as bison and wildebeest, and fish such as salmon.

The mechanisms that trigger migration vary between species but factors such as **changing day length** or **temperature**, changes in **food availability** and **genetic factors** are all thought to be important.

The table identifies some methods used to track migration of animals.

Method	Description
Direct observation	This simply involves visual identification of the species and estimation of numbers. Data can be compiled from along the migration route. This method is limited by the need to actually see the animal and is, therefore, largely restricted to daytime.
Radio tracking	A transmitter that emits radio waves (beep sounds) is attached to the animal (for example, on a collar). A VHF receiver picks up the signal, allowing the scientists to follow the migration. This is limited by both the size of the transmitter and the need for the scientist to be within range to receive the signal.
Satellite tracking	Signals from a transmitter that are attached to the animal are received by networks of orbiting satellites and bounced back to receiving stations on Earth. Information from the satellites allows the location of the animal to be identified and the migratory path to be followed remotely.
Marking	The animal is captured and an identification code attached. This information can be on a metal band placed around a bird's leg (**ringing** or **banding**), or on a metal tag that can be attached to an animal's body (**tagging**). On recapture, information on weight, general health and location can be compiled. Very large numbers of animals need to be recaptured to gain meaningful information about the migratory route.
Radar	Pulses of electromagnetic waves are emitted from an antenna. When these hit a group of animals, some of the waves are bounced back to the antenna, producing an echo that allows location and distance to be calculated. As with satellite tracking, the scientist can operate the radar from a remote position, storing electronic data for collection at a later date.
Sonar	Sensors detect underwater sounds in much the same way as a bat detects its prey. ■ Passive sonar: a sensor detects the sounds being made by marine animals and uses these to identify the species. ■ Active sonar: an emitter releases a pulse of sound waves which bounces off underwater objects and animals, the echo which is produced being detected by a sensor. The distance to the object and its direction of movement can then be calculated.

Innate and learned behaviour in migration

Innate behaviour is inherited and is carried out by all individuals in a species. It is triggered by an **external stimulus**, such as a temperature change or a change in photoperiod, and is **inflexible**.

Learned behaviour develops as a result of **experience** – through watching other individuals or by trial and error. It is **flexible**.

Migration is influenced by both innate and learned aspects of behaviour. The timing and direction of the migration are thought to be innately determined, while stopover feeding sites and alterations to the route are influenced by experience gained on previous migrations; these are, therefore, learned behaviours.

EXTREMOPHILES

Extremophiles are organisms (mostly microorganisms belonging to the domain archaea) that are found living in extreme environments in which other organisms could not survive. This includes areas with high or low temperature or pH, low oxygen concentrations or high pressure. The enzymes that allow extremophiles to survive in their hostile environments are of interest to scientists as they continue to function under severe conditions and have potential uses in many industrial processes. You should be familiar with one such enzyme called TAQ polymerase, a heat-tolerant DNA polymerase which is used in PCR. TAQ polymerase is produced by thermophilic bacteria which are able to survive the high temperatures found in hot springs.

 **DON'T FORGET**

To generate ATP, some extremophiles are able to remove high-energy electrons from inorganic molecules.

 THINGS TO DO AND THINK ABOUT

For each of the following species, suggest which method of tracking would be most appropriate. Explain your answers.

1   grey whale
2   zebra
3   monarch butterfly
4   Arctic tern
5   salmon

 ONLINE TEST

How well have you learned this topic? Take the test at www.brightredbooks.net

Rufous hummingbirds migrate across the USA between their breeding grounds in the Pacific Northwest and Mexico where they spend the winter. Migration is an energy demanding activity. Through the migration the birds replenish their fat reserves by stopping off to feed on nectar from flower meadows. At night, the hummingbirds undergo torpor as a way of saving energy, allowing their fat reserves to build up faster. Scientists have found that birds in captivity mimic this by spending more time in torpor during the autumn migration period.

MICROORGANISMS

Microorganisms can be divided into prokaryotes and eukaryotes.

Prokaryotes

Bacteria do not have a membrane bound nucleus.

Archaea look similar to bacteria when viewed using a microscope, again lacking a true nucleus. However, the DNA sequences of organisms in this group are different from those of bacteria. The membranes and metabolisms of archaea are specialised to allow organisms of this kingdom to survive in extreme environments, such as thermal vents (with temperatures >100°C), extreme saline water and extremes of pH. Many produce methane gas, being found in the digestive systems of mammals such as cows and humans.

Eukaryotes

Algae are able to photosynthesise and are found in environments with available light and moisture.

Protozoa do not photosynthesise. Most are motile and catch other microbes as their food source.

Fungi (e.g. yeast) do not photosynthesise. Usually non-motile, they absorb nutrients directly from the environment.

CULTURE CONDITIONS

Growth of microorganisms in the laboratory requires:

- a water-based **growth medium** that supplies:
  (a) raw materials for synthesis of molecules. Some microorganisms are able to produce all of the complex molecules that they need, such as amino acids for protein synthesis. For other microorganisms, complex compounds such as vitamins and fatty acids must be supplied in the growth medium.
  (b) a source of energy (if the microorganism cannot use light for photosynthesis).

- a suitable environment with appropriate temperature, pH and oxygen levels.

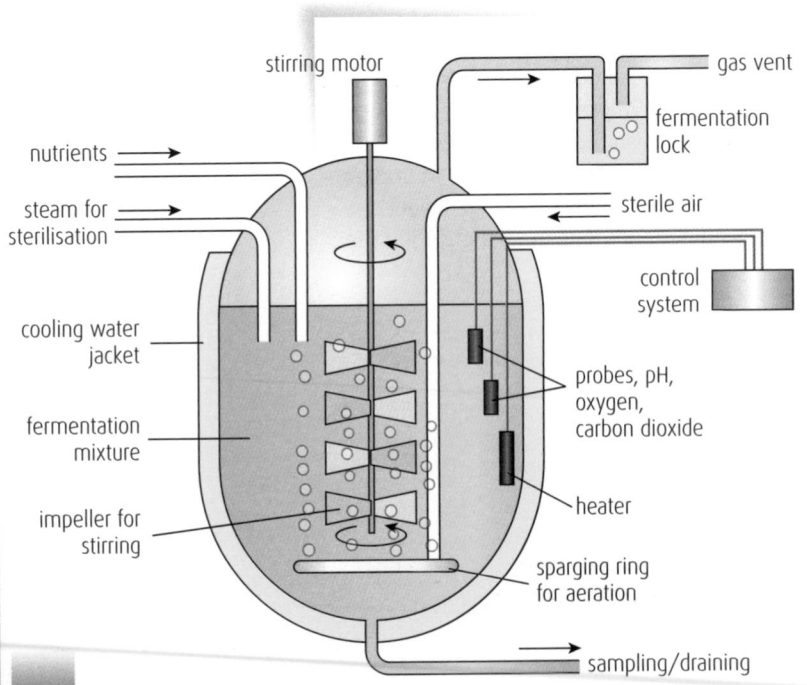

GROWTH MEDIA

Growth media can be used in either a liquid or solid form.

Liquid media

Liquid medium can be used in Petri dishes or flasks and is the growth medium used in most industrial **fermenter vessels**. The diagram on the left shows a typical fermenter. Here, the growth medium (a **broth**) can be stirred, allowing nutrients and heat to be distributed evenly and aerobic conditions to be maintained (if appropriate). The composition of the growth medium can also be monitored and altered by the addition of substances; the growth medium can be withdrawn from the fermenter vessel for retrieval of useful products.

contd

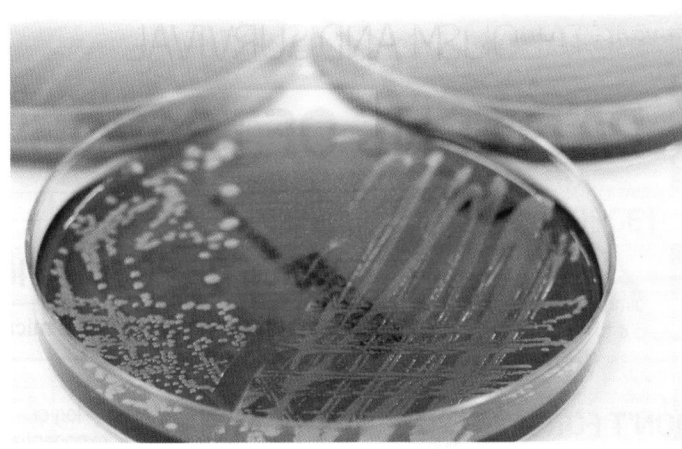

Solid media

Solid medium is produced by the addition of **agar** (a jelly-like substance) to a liquid medium, causing the liquid to solidify. Most culture work carried out in schools uses a solid medium (poured into Petri dishes while molten).

Media are also classified by the composition of nutrients that they contain.

Complex media

The exact chemical composition of nutrients in this media is 'undefined' or unknown, because the source of amino acids is a complex meat or yeast extract. Complex media set with agar in Petri dishes is used routinely for the culture of microbes in schools.

Defined media

The chemical composition of the medium is known and is in a relatively pure form. Some genetic studies and nutritional studies require defined media.

 **DON'T FORGET**

Bacteria plated up on agar plates require oxygen, so the plates must not be sealed completely.

ENVIRONMENT

In order to produce the maximum yield of product, it is important to maintain optimum environmental conditions.

Sterility	**Aseptic techniques** are used to eliminate any unwanted, contaminating microorganisms that would affect the growth of the desired microorganism. All equipment and culture media must be sterilised before use, either by heating under pressure in an **autoclave** (a pressure cooker) or, in the case of large industrial fermenter vessels, steam cleaned for 30–60 minutes. During microbial culture, the aseptic techniques shown here are used.

Pass neck of culture bottle through flame (keep lid in hand)

Pass neck of culture bottle through flame and replace lid

Sterilise loop in Bunsen flame

Streak a third and forth time, sterilising the loop each time. Then seal plates and incubate

Sterilise loop in Bunsen flame

Carefully touch single colony with loop

Streak an area of agar on labelled Petri dish

Streak across previous streaks and sterilise loop again

Temperature	A suitable temperature of 30–37°C provides the optimum temperature for growth of bacterial cultures. Fungi, however require lower temperatures (25–27°C).
Oxygen	Filtered air is bubbled through liquid media (**aeration**) to provide oxygen for aerobic microorganisms in fermenter vessels.
pH	The optimum pH of the culture medium is maintained through the addition of buffers or through the addition of acids and alkalis. Most bacteria grow better in media of pH7. Fungi usually prefer pH5–6.

THINGS TO DO AND THINK ABOUT

1   When culturing microbes in a school laboratory, it is important to avoid growing pathogens. This can be done by ensuring that growth occurs in the correct conditions. Explain.

 ONLINE TEST

Once you've learned about this topic, test your knowledge at www.brightredbooks.net

GENETIC CONTROL OF METABOLISM

Wild strains of microorganisms can be changed to provide large quantities of products for use by humans. Techniques used include:

- mutagenesis
- selective breeding and culture
- recombinant DNA technology

MUTAGENESIS

Mutagenic agents, such as **ultraviolet (UV) light**, **X-rays** and γ-rays, and **mutagenic chemicals**, such as **mustard gas**, can cause damage to DNA, resulting in changes to the base sequence of a gene (a **mutation**). As a result, the cell stops making the protein that would have been coded for by the normal gene. If the protein is involved in a metabolic pathway, the pathway may be blocked and **intermediate metabolites** may build up. Where these are of use to humans, the intermediate metabolite can be harvested as a **product**. Microorganisms that are used in industrial processes are often mutant strains that have had their base sequences deliberately altered at specific sites. The mutant microorganisms tend to be **genetically unstable** and, with continued cell divisions, the DNA base sequence can revert back to the normal type. It is, therefore, important that the cultures are carefully monitored to ensure that the mutant type is selected for use.

SELECTIVE BREEDING

Wild strains of the microorganism are grown in culture under optimal conditions. Pure strains for the required task can then be isolated and selected for subculture.

RECOMBINANT DNA TECHNOLOGY

A gene from one organism is transferred into a microorganism, so that the microorganism produces plant or animal proteins. Scientists use **plasmids** and **artificial chromosomes** as **vectors**, to transfer genes coding for proteins that are useful to humans into microorganisms.

Producing recombinant plasmids

Plasmids are small circular pieces of DNA – carrying between 25 and 30 genes – that are separate from the chromosomes in a cell. They are found in most bacteria and some fungi. The diagram shows the process used to insert a human insulin gene into a plasmid.

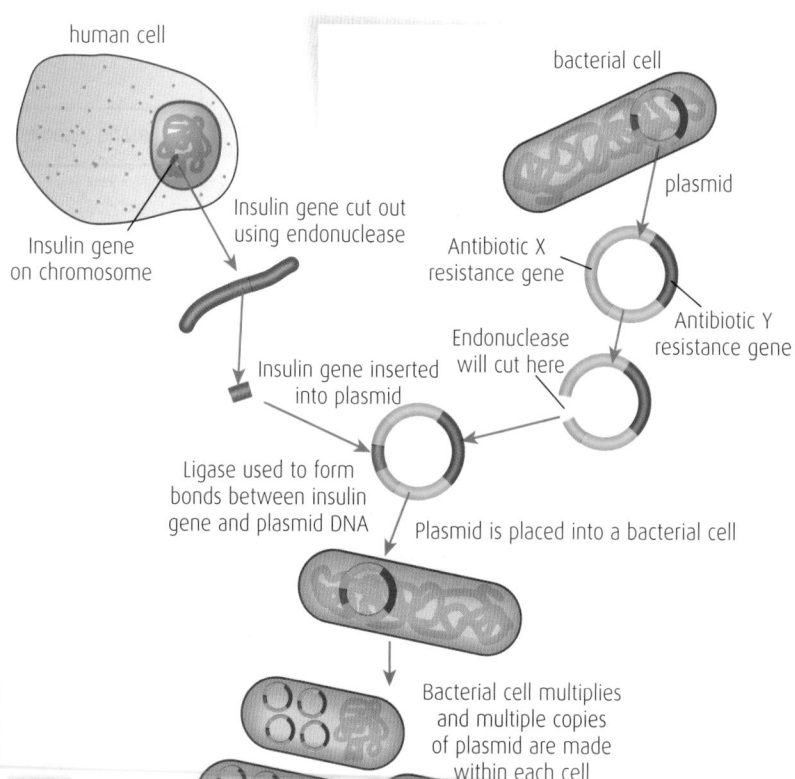

human cell

Insulin gene on chromosome

Insulin gene cut out using endonuclease

bacterial cell

plasmid

Antibiotic X resistance gene

Antibiotic Y resistance gene

Endonuclease will cut here

Insulin gene inserted into plasmid

Ligase used to form bonds between insulin gene and plasmid DNA

Plasmid is placed into a bacterial cell

Bacterial cell multiplies and multiple copies of plasmid are made within each cell

contd

Usually the **recombinant plasmid** will have several **marker genes** inserted into its DNA. These allow cells that have taken up the plasmids to be identified. Marker genes either code for **fluorescent proteins** that can be seen through a microscope, or give the cell **antibiotic resistance**.

Additional genes for **self-replication** and **regulatory sequences** that allow control of gene expression are also inserted into the plasmid. These will cause multiple copies of the plasmid to be made within the cell, increasing the quantity (yield) of product that can be harvested from the culture. Other genes may be inserted as a **safety mechanism** to make sure that the microorganism cannot survive in an external environment.

Restriction endonucleases and DNA ligases

Restriction endonucleases are enzymes that recognise specific nucleotide sequences (**restriction sites**) on a DNA molecule and cut the DNA into fragments by breaking bonds at specific points along the nucleotide sequence. If the cut goes straight across the DNA molecule, the two strands of nucleotides will be cut at the same place producing **blunt ends**. However, if the nucleotides are cut at different points, several nucleotides apart, each fragment ends with a short single stranded segment (**a sticky end**). As one restriction enzyme will always produce DNA fragments with the same sticky ends, fragments from two different DNA molecules cut using the same enzyme can be brought together, forming hydrogen bonds between base pairs on the complementary sticky ends. Once in position, **DNA ligase** is used to form new bonds, combining the two types of DNA more permanently.

DON'T FORGET

Sometimes polypeptides produced by recombinant bacteria are folded incorrectly or lack post translational modifications. The use of recombinant yeast cells avoids this.

DON'T FORGET

Restriction endonucleases act like scissors, cutting DNA into fragments, and DNA ligases act like glue, sticking DNA fragments together.

ETHICAL CONSIDERATIONS IN THE USE OF MICROORGANISMS

The use of biotechnology has raised ethical issues with public-interest groups arguing that the long-term risks outweigh the benefits.

Argument against genetic engineering	Counter argument
Microorganisms carrying a dangerous gene may escape or be stolen.	Manufacturers carry out detailed risk assessments and maintain suitable standards to ensure that manufacturing processes and products are safe.
Genetically modified organisms or gene sequences should not be the property of the companies that develop them, but should be available to all.	It is expensive to develop the genetically modified microorganisms. If patents cannot be taken out, companies will not invest money to develop this technology.

 THINGS TO DO AND THINK ABOUT

1   The flow chart below represents the programming of *E. Coli* bacteria to produce human insulin.

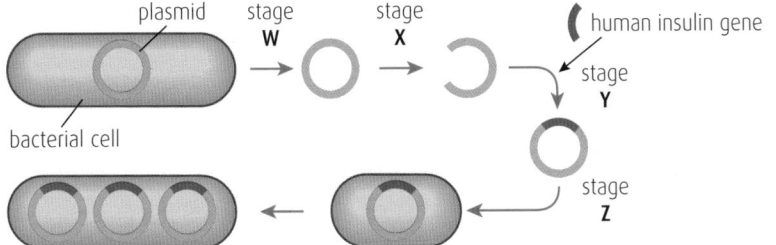

Which line in the table below identifies correctly the stages at which an endonuclease and a ligase are used?

	Endonuclease	Ligase
A	Stage X	Stage W
B	Stage Y	Stage Z
C	Stage X	Stage Y
D	Stage Y	Stage X

ONLINE TEST

Head online and test yourself on this at www.brightredbooks.net

THE SCIENCE OF FOOD PRODUCTION

ONLINE

Watch this clip about food security at www.brightredbooks.net

ONLINE

Read about this global issue at www.brightredbooks.net

ONLINE

Read about world food problems at www.brightredbooks.net

FOOD SECURITY

Food security is an important global issue currently facing the human population. Food security is the ability of human populations to access food in sufficient **quantity** and of sufficient **quality**.

Humans must be able to provide enough food to meet their needs. This food must be **safe** and **nutritious**, providing the correct balance of nutrients for good health. The world is facing a potential food security crisis because the human population has been increasing rapidly, having reached 7 billion in 2012 and being predicted to reach over 9 billion by the year 2050!

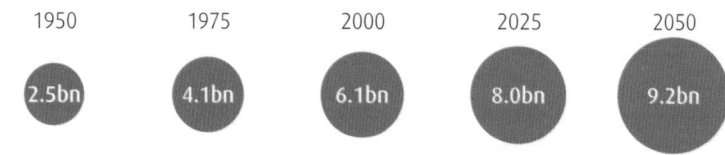

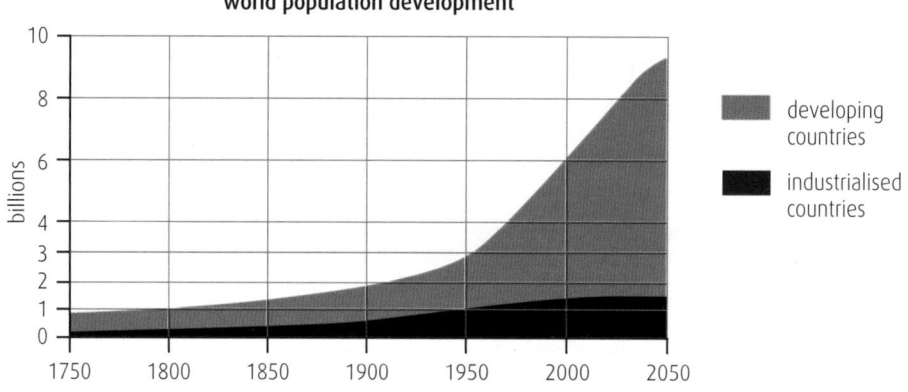

Several natural factors which affect food production are:

- drought
- pests
- flooding
- disease
- desertification
- resistance to insecticides

It is important that food production is **sustainable** and does not degrade the natural resources that are vital for agriculture. Humans need to be able to provide enough quality food without damaging the environment through:

1 global warming – increased food production causes an increase in the greenhouse gases that contribute to global warming

2 pollution – pesticides and fertilisers can pollute land or nearby rivers

3 deforestation – forests may be cleared to make way for agricultural land, contributing to the greenhouse effect and also leading to habitat loss which reduces species diversity

4 soil erosion – intensive farming practices can cause loss of fertile top soil by erosion

5 reducing soil fertility – harvesting crops means nutrients are not naturally recycled and so soil fertility is reduced.

DON'T FORGET

Photosynthesis is the process by which green plants (autotrophs) produce food using light energy. The light energy is trapped by the pigment chlorophyll in chloroplasts.

AGRICULTURAL PRODUCTION

The production of all food depends on photosynthesis. Animals, including humans, must eat plants (and other animals) to obtain energy. Most human food comes from a small number of plant crops, including cereals, potatoes, root vegetables and legumes.

Factors which limit plant growth

Light intensity	Light intensity limits the rate of photosynthesis. As light intensity increases, the rate of photosynthesis increases and so does plant growth.
Carbon dioxide concentration	Carbon dioxide concentration also limits the rate of photosynthesis. An increase in CO_2 concentration increases the rate of photosynthesis and so increases plant growth.
Temperature	At low temperatures plant growth is limited. This is because the reactions of photosynthesis are controlled by enzymes with an optimum temperature of around 35°C. As temperature increases up to 35°C, plant growth increases. Above the optimum temperature, plant growth rate decreases as enzymes are denatured.
Availability of nutrients	Green plants require several chemical elements for healthy growth. If these are not present in the soil, growth rate will be affected.
Pests and disease	Plants can be damaged by inverteberate pests or their growth can be inhibited by bacterial and fungal diseases.
Competition	Plant growth may be inhibited if there is competition with other plants for light, water or nutrients in the soil.

If there is limited land available to grow crops to meet the demand for food, then all of the factors which limit plant growth have to be controlled and farming becomes intensive.

Intensive Farming Practices

Practice	Effect
Growth of high yield crops (cultivars)	A high yielding variety of a crop plant is created or selected and maintained through cultivation
Fertilisers	These give increased crop yield
Pesticides	Crops are protected from pests, diseases and competition by use of insecticide, fungicide or herbicide

Intensive farming results in an increase in food production but may have undesirable side effects, for example herbicides and pesticides reduce biodiversity.

Energy availability

In any food web there is significant energy loss from each trophic level, mainly as heat energy due to metabolic processes. Therefore, only a small proportion of energy is available for the next level. As a result, if plants are used for food, they provide more energy per unit area than livestock animals.

Some habitats are unsuitable for crop production. However, livestock production may be possible in managed or wild habitats unsuitable for crops, for example hill sheep farming.

Cereals

Potatoes

Root vegetables

Legumes

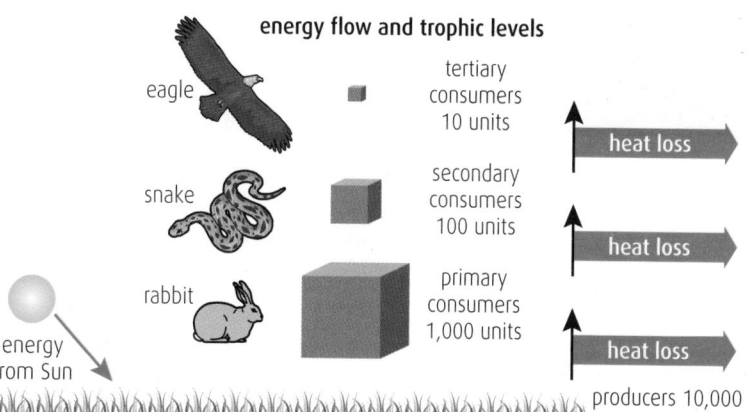

energy flow and trophic levels

eagle — tertiary consumers 10 units — heat loss

snake — secondary consumers 100 units — heat loss

rabbit — primary consumers 1,000 units — heat loss

energy from Sun

producers 10,000 units of energy

THINGS TO DO AND THINK ABOUT

1 Describe two ways in which food production from crops can be increased.

2 Explain why crop plants provide more energy per unit area than livestock

3 Describe three ways in which energy is lost from animals.

4 List three habitats which would be unsuitable for growing crops.

 ONLINE TEST

How well have you learned this topic? Take the test at www.brightredbooks.net

PLANT GROWTH AND PRODUCTIVITY 1

DON'T FORGET

Light energy which is not absorbed by plant pigments is either reflected or transmitted.

ONLINE

Find out about leaf pigments and absorption spectra at www.brightredbooks.net

PHOTOSYNTHESIS – LEAF PIGMENTS

Green plants use light energy from the Sun to produce carbohydrates. Light energy is absorbed by plant pigments in the grana of chloroplasts.

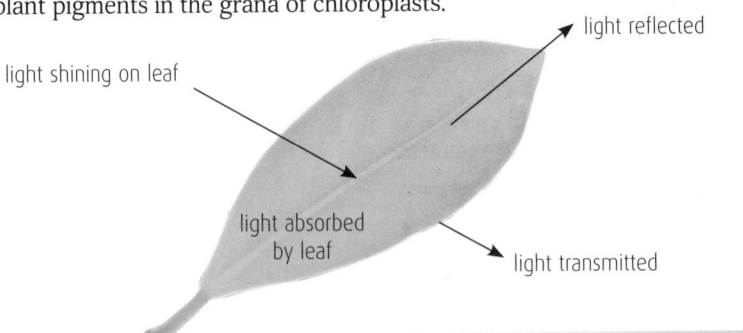

THE SPECTRUM OF VISIBLE LIGHT

The light we see is known as white light. It is a combination of all visible colours of the electromagnetic spectrum. If white light is passed through a glass prism, it is separated into its different wavelengths seen as light of different colours.

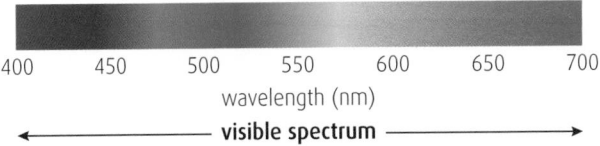

If white light is passed through a sample of leaf pigments, then through a glass prism, the following absorption spectrum is produced.

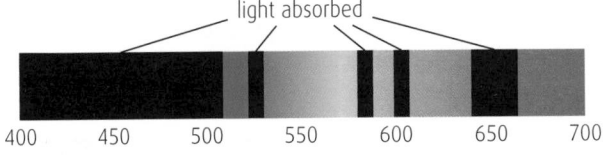

ABSORPTION SPECTRA FOR LEAF PIGMENTS

Different plant pigments absorb different wavelengths (colours) of light. The ability of these pigments to absorb different wavelengths can be measured using a spectrophotometer and a graph of results is produced. This graph is an **absorption spectrum**.

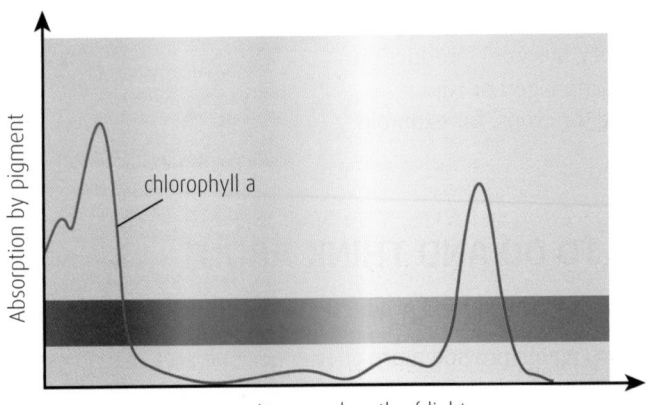

Chlorophyll a

Chlorophyll a absorbs red and blue light.

contd

Chlorophyll b and the carotenoids

The accessory pigments, chlorophyll b and the carotenoids (carotene and xanthophyll), absorb light of different wavelengths to chlorophyll a. They pass energy to chlorophyll a. This extends the range of wavelengths that can be absorbed. Therefore, more light energy is available for photosynthesis.

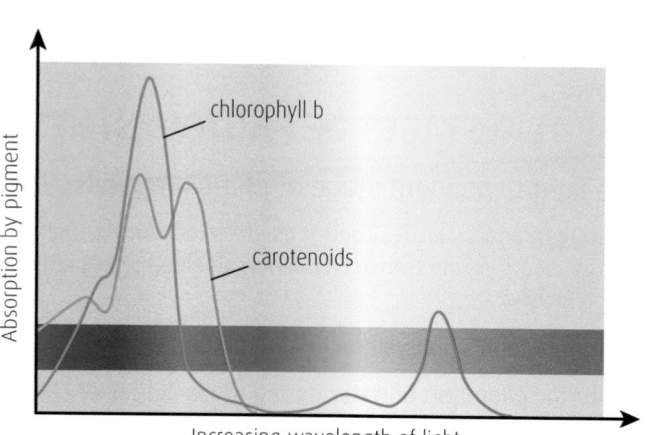

ACTION SPECTRUM

An action spectrum is a graph which shows the rate of photosynthesis at different wavelengths of light.

Action spectrum for leaf pigments

The action spectrum has a similar shape to the combined absorption spectra for the plant pigments. This is evidence that the plant pigments absorb light for use in photosynthesis.

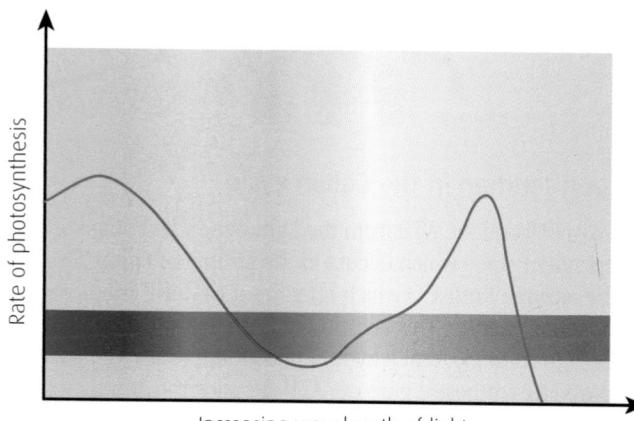

Pigment	Function
chlorophyll a	This absorbs red and blue light, and is the only pigment that can participate directly in light reactions
chlorophyll b	These are accessory pigments which absorb other wavelengths of light. Captured energy is passed on to chlorophyll a. The accessory pigments are important as they increase the energy available for photosynthesis.
xanthophyll	
carotene	

THINGS TO DO AND THINK ABOUT

1 Describe what the black bands represent in an absorption spectrum.

2 Explain why chlorophyll appears green.

3 What does an action spectrum for photosynthesis represent?

4 Explain why it is beneficial for a plant to have more than one plant pigment.

 ONLINE TEST

Once you've learned about this topic, test your knowledge at www.brightredbooks.net

PLANT GROWTH AND PRODUCTIVITY 2

DON'T FORGET

Photosynthesis occurs
in two stages: a light
dependent stage and the
Calvin Cycle.

DON'T FORGET

Both hydrogen and ATP
are used in the next
stage of photosynthesis,
carbon fixation.

THE TWO STAGES OF PHOTOSYNTHESIS

The light dependent stage of photosynthesis

When light energy is absorbed, it excites electrons in the leaf pigments and these high
energy electrons are transferred along an electron transfer chain. This releases energy
which is used:

1 for photophosphorylation of ADP to form ATP by the enzyme ATP synthase

2 for the photolysis of water, during which light energy splits water molecules into
hydrogen and oxygen.

Hydrogen binds to the coenzyme NADP to form NADPH. Oxygen is released as a
by-product of photosynthesis. These events take place in the grana of chloroplasts.

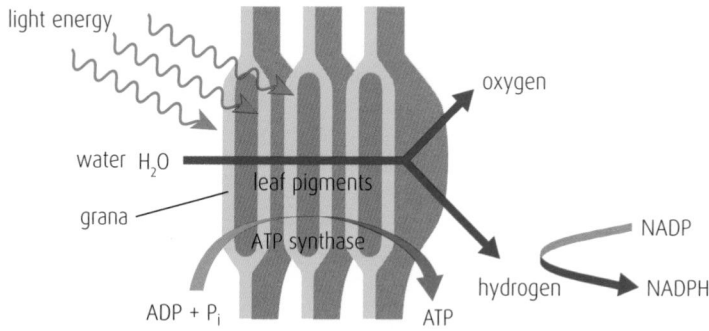

Carbon fixation in the Calvin cycle

The NADPH and the ATP from the light dependent stage are required for the light
independent stage which occurs in the stroma of chloroplasts. Carbon dioxide is fixed
by the enzyme RuBisCO which attaches it to RuBP (ribulose bisphosphate) to form
3-phosphoglycerate. This compound is phosphorylated by ATP and joins with hydrogen from
NADPH to form glyceraldehyde-3-phosphate (G3P). G3P is used to regenerate RuBP and is
also used to synthesise glucose. ATP supplies the energy for the reactions to take place.

ONLINE

The link at
www.brightredbooks.net
explains
photophosphorylation and
the synthesis of ATP.

ONLINE

Try the quiz about the
Calvin cycle at
www.brightredbooks.net

ONLINE

Watch the Calvin
cycle animations at
www.brightredbooks.net

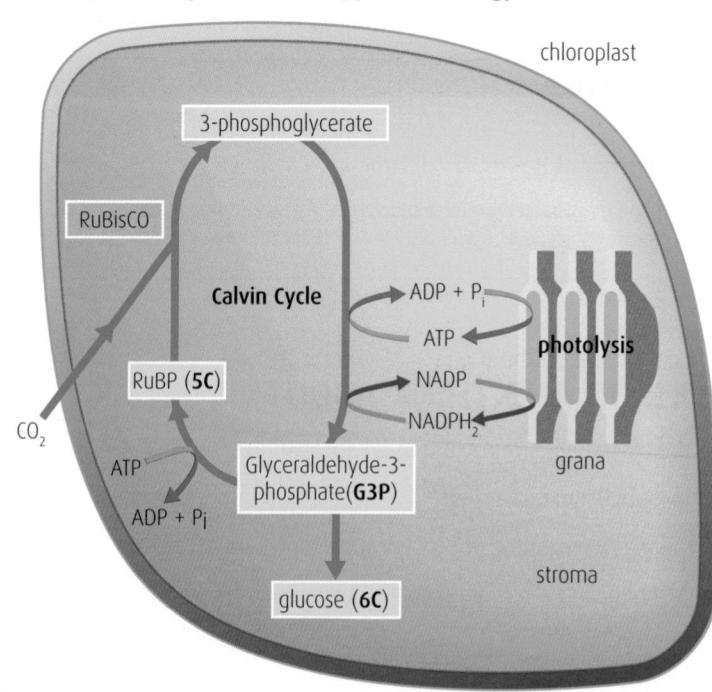

contd

There are several possible fates for the glucose produced in the Calvin cycle:

- Some will be converted into starch (a storage carbohydrate).
- Some will be converted into cellulose for cell walls (a structural carbohydrate).
- Some is used to produce energy in respiration.
- Some passes to other biosynthetic pathways.

Plant productivity

Plant productivity is the rate of production of new plant material (biomass) **per unit area per unit of time** (in grams per square metre per year). **Assimilation** is the increase in mass of a plant due to its production of organic molecules (sugars and starch) during photosynthesis. However, carbohydrates are also used up in respiration to produce energy. Therefore, **net assimilation** is the increase in mass due to photosynthesis minus the loss in mass during respiration. Net assimilation can be calculated by measuring the increase in dry mass per unit area of a leaf. The **biological yield** of a crop is the total biomass produced, whereas the **economic yield** is the mass of the desired product.

The **Harvest Index = dry mass of economic yield** (e.g. grain)/**dry mass of biological yield** (e.g. grain, straw and roots)

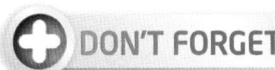

DON'T FORGET

Remember that several factors limit the rate of photosynthesis, including **light intensity, carbon dioxide concentration** and **temperature**.

> **Example:**
> Calculate the harvest if dry mass of biological yield is 20 000 kg and dry mass of economic yield is 7000 kg.

> **Answer:**
> Harvest Index = dry mass of economic yield/dry mass of biological yield
> Harvest index = 7000/20 000 = 0·35

THINGS TO DO AND THINK ABOUT

You may have carried out an experiment to investigate photolysis using the Hill Reaction. You should be able to explain that DCPIP changes from blue to colourless when it acts as a hydrogen acceptor and that during photolysis hydrogen is released from water and so DCPIP becomes colourless. The rate of decolourisation of DCPIP in the Hill Reaction is a measure of the rate of the light dependent stage of photosynthesis.

1. Look at the diagram of the Calvin cycle and the position of the enzyme RuBisCO. Explain what would happen to the concentration of carbon dioxide inside the leaf on a hot day when stomata close to conserve water. What would happen to the concentration of oxygen?

   Hint: The normal activity of RuBisCO will be inhibited in this situation.

2. Explain the meaning of each of these terms: (a) plant productivity (b) net assimilation (c) biological yield (d) economic yield.

3. You may have carried out experiments to investigate the factors which limit the rate of photosynthesis. Draw graphs showing the effect of temperature, light intensity and carbon dioxide concentration on the rate of photosynthesis.

4. Describe an experiment which could be carried out to compare the net assimilation rate of the same plant in different concentrations of carbon dioxide.

ONLINE TEST

Head online and test yourself on this at www.brightredbooks.net

PLANT AND ANIMAL BREEDING 1

VARIATION

Plant and animal breeders try to develop new and improved varieties of crops and livestock to provide **sustainable food sources**. They manipulate or change the heredity (genetics) of the organism.

Single gene inheritance controls characteristics which show discrete variation. Characteristics which show continuous variation are controlled by several genes. This is **polygenic inheritance**. Characteristics which are controlled by polygenic inheritance can also be influenced by the environment.

Reason for altering an organism's heredity	Benefit	Example
increase in yield	more food produced	barley
increase in nutritional value	increase in protein content	rice
resistance to disease	crop not damaged by disease	blight-resistant potatoes
resistance to pests	crop resistant to insect, fungus or worm	soya bean resistance to nematode worm
to survive a particular environment	adapted to grow in hot dry environment	corn
to make it more suitable for rearing or harvesting	a more uniform crop height makes harvesting easier, increasing yield	wheat

CULTIVARS

A **cultivar** is a plant that has been created or selected intentionally for desirable characteristics that can be maintained by **cultivation**. A cultivar is different from others in at least one characteristic or trait. Agricultural food crops are almost exclusively cultivars.

FIELD TRIALS

A field trial is an experimental investigation in an organism's natural environment (rather than in the laboratory.) **Plant field trials** are used to see whether or not a particular treatment makes a difference to the crop. Measurements such as height, grain size, yield or incidence of disease can be used to determine the effects of a treatment. Trials can be used to evaluate:

* the performance of different cultivars in a range of environments
* the effects of different treatments such as pesticides or fertilisers
* GM crops.

contd

ONLINE

Read about how to conduct a plant field trial at www.brightredbooks.net

ONLINE

Find out about maize field trials at www.brightredbooks.net

The design of field trials

Reliable and valid field trials require:

1 **Careful selection of treatment** – if the treatment to be investigated is the effect of fertiliser, the treated plot would be given fertiliser and a control plot would have no fertiliser. All other variables would be kept constant. The treated and untreated plots should be as similar as possible (same soil, moisture and slope) to allow a valid comparison of treated and untreated areas.

2 **Replicates** – use several trial plots to increase **reliability** of results. This reduces the effect of variability within samples.

3 **Randomisation** – the treated and control areas should be scattered randomly across the site of the trial to eliminate the possibility of bias when measuring the effect of the treatment.

ONLINE TEST

How well have you learned this topic? Take the test at www.brightredbooks.net

ONLINE

Investigate current research which is being carried out in Scotland to gain a greater understanding of potato blight using the link at www.brightredbooks.net

ONLINE

Investigate research into various soft fruit cultivars in Scotland using the link at www.brightredbooks.net

 THINGS TO DO AND THINK ABOUT

1 Draw a table showing five examples each of **discrete** and **continuous** variation. Give examples from both plants and animals.

2 Design a plant field trial to compare the effectiveness of two pesticides on a particular variety of wheat.

Late blight in potatoes is caused by the fungus *Phytophthora infestans*. This infects leaves, stems and tubers and can cause devastating crop losses, such as in the Irish Potato Famine (1845–1852). Changes in the blight fungus now mean that almost all the varieties that previously had high or moderate resistance are affected.

PLANT AND ANIMAL BREEDING 2

DON'T FORGET

An organism's genotype is **homozygous** when it has two identical alleles of a gene. It is **true breeding**. When it has two different alleles of a gene it is **heterozygous**.

SELECTIVE BREEDING

Plant and animal breeders *choose* the parent organisms with the desired characteristics and use them in their breeding programmes. The aim is to produce offspring with the combined characteristics of the two parents.

Outbreeding is the mating of unrelated or distantly related members of a species. Animals and cross-pollinating plants are naturally outbreeding.

Inbreeding is the mating of closely related individuals. This can happen naturally, for example in self-pollinating plants. In inbreeding programmes, **selected** plants or animals are bred for several generations until the population breeds true to the desired type due to the elimination of heterozygotes.

Inbreeding depression

Inbreeding results in **homozygosity** leading to the accumulation of harmful (**deleterious**) homozygous alleles and increasing the chances of offspring being affected by recessive traits – this is **inbreeding depression**. Inbreeding depression can result in loss of vigour and poor general health, reduced size, fertility or yield. Self-pollinating plants are less susceptible to inbreeding depression since **natural selection** eliminates the deleterious alleles over many generations. Breeders try to avoid inbreeding depression in naturally outbreeding species by choosing parents with the *desired* characteristic but which are otherwise genetically diverse.

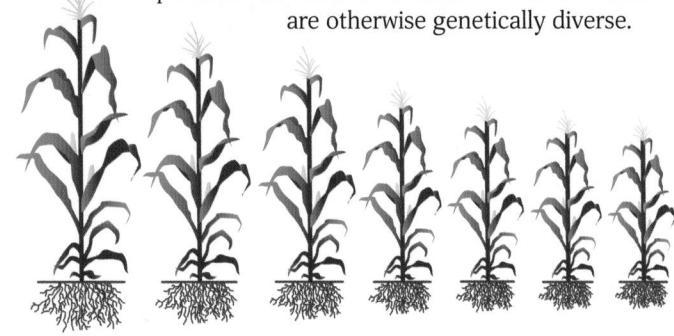

Inbreeding depression in the white tiger and hybrid corn

CROSSBREEDING AND F1 HYBRIDS

ONLINE

Read more about crossbreeding in beef cattle at www.brightredbooks.net

ONLINE

Find out about the many varieties of cross bred sheep at www.brightredbooks.net

In order to prevent inbreeding depression and to introduce new alleles to plant or animal lines, **crossbreeding** is carried out. This is when a plant or animal is crossed with another organism with a different but desired genotype, creating offspring with traits from both parents, improving characteristics and producing organisms with **hybrid vigour**.

An **F1 hybrid** is the offspring of such a cross. In plant breeding, an F1 hybrid is the cultivar derived from two different parent cultivars. For example, almost all modern cultivars of edible (seedless) bananas are hybrids of two wild banana species with seeds.

In animals, a **crossbreed** is the result of breeding two different purebred parents of different breeds. In cattle, purebred females adapted to a specific environment can be crossed with purebred bulls from another environment to produce a generation with traits of both parents. Highland cattle have a long history as a pure breed and are used to improve hardiness when crossbred with other breeds. The Scotch Mule breed of sheep is a crossbreed of a Scottish Blackface ewe and a Bluefaced Leicester Ram. The ram passes many desirable traits to his crossbred daughters, which have improved maternal qualities including milk production and early maturity.

contd

In both plants and animals, crossbreeding results in an F2 generation with a wide range of genotypes. Therefore, if the breeder wishes to maintain the new desired breed a process of **selection** and **backcrossing** is necessary. However, it may also be useful for the breeder to maintain the two parent breeds to produce F1 crossbred animals.

Mutation breeding in plants involves exposing seeds to chemicals or radiation in order to generate mutants with desirable traits to breed with other cultivars. Mutation breeding has improved certain crops, giving disease resistance, dwarf habit or a change in the chemical or nutritional plant composition.

DON'T FORGET

A **mutation** is a change in the genetic information in a cell. Mutations occur randomly within a population and introduce variation to the species. Mutagenic **agents**, such as UV-light, X-rays and the chemical mustard gas, cause the rate of mutation to increase.

ONLINE

Find out how mutation breeding has improved rape seed plants at www.brightredbooks.net

DON'T FORGET

The **phenotype** of an organism is its physical appearance, for example yellow seeds or green seeds. The **genotype** is the genes or alleles an organism has for a particular characteristic, for example YY, Yy or yy.

TEST CROSS

A test cross is used to determine the genotype of an individual with a **dominant phenotype** to find out if it is **homozygous** or **heterozygous**. The organism with the dominant phenotype is crossed with an organism that is **homozygous recessive** for the same characteristic. If the organism with the dominant phenotype is a homozygote, then all F1 offspring will show the dominant phenotype. If the organism with the dominant phenotype is a heterozygote, the F1 offspring will show a 1:1 ratio of heterozygotes and recessive homozygotes. In the example shown, yellow seed colour is dominant to green. The test cross is carried out with a plant with green seeds, the homozygous recessive (yy). The two possible results of the test cross are shown in the diagram.

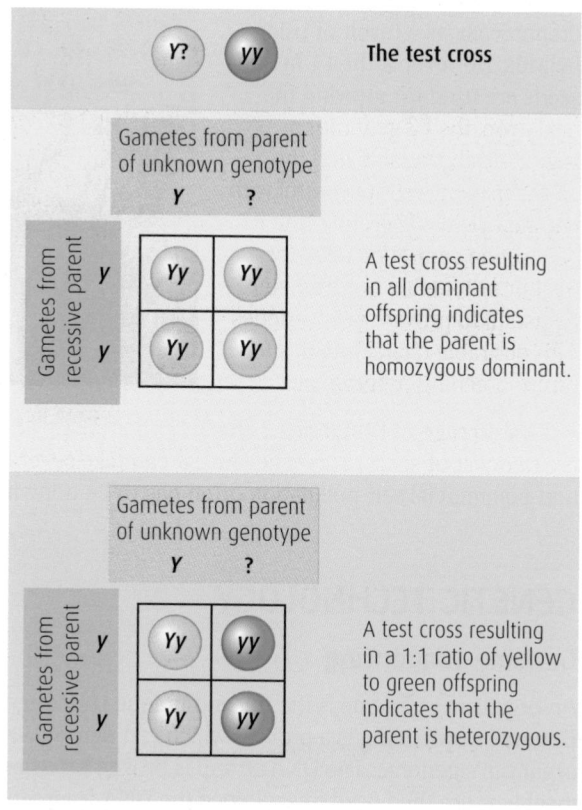

A test cross can be used in breeding programmes in both plants and animals in order to identify unwanted individuals with heterozygous recessive alleles.

 THINGS TO DO AND THINK ABOUT

In the seeds of traditional rapeseed, the level of erucic acid, which is nutritionally undesirable (causing heart damage in animal tests), is higher than that of other oil seed crops. However, mutation breeding has resulted in a 'zero erucic acid' variety.

Mutant genes have been successfully introduced into commercial crops such as rice, wheat, maize, barley and soybean that significantly enhance their nutritional value.

ONLINE TEST

Once you've learned about this topic, test your knowledge at www.brightredbooks.net

PLANT AND ANIMAL BREEDING 3

F1 HYBRIDS IN PLANTS

When plant breeders identify a desired characteristic in a plant, they can self-pollinate this plant over several generations until, each time the seeds are germinated, identical plants are produced. This is an **inbred line** and can take many years to achieve. The breeder can cross-pollinate stable inbred lines of two **different** plants with **different** desired characteristics by hand. Plants grown from the seeds produced in this way should have the combined traits of the two parents. These are **F1 hybrids**. The crop which is produced is heterozygous and relatively uniform, usually with increased vigour, disease resistance and yield (hybrid vigour). The photographs show a hybrid wheat variety tolerating waterlogged soil better than the control.

F1 hybrid seeds have had a positive impact on food production in agriculture; yields of corn in the USA and rice yields in China have both increased dramatically as a result of using hybrids. However, if the F1 hybrid seeds are used for growing the next crop, the F2 generation is genetically variable and the plants may be less successful with reduced yields. Therefore, the F2 generation cannot be used for further production but may still be useful to provide new varieties with desirable characteristics for future breeding programmes.

Note that the control is not an F1 hybrid

A disadvantage of F1 hybrids is the greater cost of seeds. This is because it can take several years to produce inbred lines and non-commercial self-pollination often has to be done manually, which is time consuming.

GENETIC TECHNOLOGY

Genome sequencing

An organism's genome is the full and unique set of genetic instructions in its DNA. **Genome sequencing** is a process which can determine the DNA sequence of an organism's genome. The DNA strand is broken up using enzymes. Gene sequencing machines are then used to sequence the DNA fragments and computer programmes are used to compare overlapping sections to recreate the sequence of the original DNA strand. Genome sequencing allows organisms with desirable genes to be identified and used in breeding programmes.

contd

Genetic transformation

Genetic transformation is the alteration of the genome of a cell by the insertion of a gene or genes from another organism, either naturally or artificially. The transformed genome can then be used in breeding programmes. Genetically modified crops are useful for:

- protection from threats such as low temperatures, insects and viruses
- increasing nutritional value or yield
- increasing tolerance of salt or drought
- reducing the need for as much fertiliser or pesticide.

The soil bacterium *Bacillus thuringiensis (Bt)* produces a protein (**Bt toxin**) which is toxic to certain insects. When eaten, the toxin is activated and destroys the insect's gut. There are different strains of Bt toxin which are effective against different insect pests. Scientists can remove the Bt toxin gene from bacteria and insert it into plants, allowing plants to produce their own toxin that offers protection against damage by insects. This removes the need to use chemical insecticides.

Glyphosate is a herbicide (weed killer) which is particularly effective against broadleaf weeds and grasses that compete with crops. It is an enzyme inhibitor in the production of amino acids and, when absorbed through plant foliage, it is transported to the growing points of the plant, killing it. Genetically engineered crops that are **glyphosate tolerant** can now be produced. They contain the glyphosate resistance gene, allowing farmers to control weeds without damaging their crop.

Rice is low in vitamin A and, so, vitamin A deficiency and its associated visual defects are prevalent in regions where rice is the staple diet. Two genes for beta carotene can be inserted into the normal rice **genome**, so that rice plants can make beta carotene, a precursor of vitamin A. The body converts beta carotene to vitamin A as required. **Golden rice** is a genetically engineered rice cultivar which contains beta carotene and is, therefore, yellow–orange in colour, as shown in the photographs.

 **DON'T FORGET**

Genetic engineering is the genetic alteration of a cell by insertion of a gene or genes from another organism. The cell has been genetically modified, producing a GM (genetically modified) organism.

 THINGS TO DO AND THINK ABOUT

1 Give the definition of (a) an inbred line, (b) an F1 hybrid (c) genetic transformation, and (d) genome sequencing.

2 Research 'golden rice' and produce a table comparing the arguments for and against its production.

3 Research the use of Bt toxin in the genetic modification of plants. Name three insect pests which can be killed by Bt toxin.

4 Some weeds are becoming resistant to glyphosphate. Describe the problems this would give a farmer using this herbicide.

 ONLINE TEST

Head online and test yourself on this at www.brightredbooks.net

CROP PROTECTION 1

DON'T FORGET

When two or more members of a community need the same resource, competition occurs. Plants compete for light, water, minerals and root space.

DON'T FORGET

Plant productivity is the rate of production of new plant material (biomass) **per unit area per unit of time**. Factors limiting the rate of photosynthesis and, therefore, plant productivity include **light intensity, carbon dioxide concentration** and **temperature**.

ONLINE

The link at www.brightredbooks.net explains about the different types of weeds and how best to control them.

ONLINE

Read about characteristics of different types of weeds and their impact at www.brightredbooks.net

CROP PLANT PRODUCTIVITY

Plant productivity can be reduced by factors including weeds, pests and disease. Weeds are plants that grow where they are not wanted and that compete with crops for space, light, water and nutrients.

Annual weeds

An **annual weed** has a short life cycle, that is it completes its life cycle (goes from seed to seed) in one growing season. Annual weeds grow rapidly, producing flowers and seeds quickly. The seeds of annual weeds are produced in large numbers and can survive in the soil for many years, until conditions are suitable for germination and growth. They can germinate at lower temperatures than many cultivated plants.

Characteristics of annual weeds:

- rapid growth
- short life cycle
- high seed output
- long-term seed viability.

Perennial weeds

Perennial weeds are persistent weeds that continue to grow every season. They usually die down in the winter and regrow each spring. They have **competitive adaptations** which allow them to compete successfully with cultivated plants. They can reproduce both sexually, by seed production, and asexually by the **vegetative** reproduction. These are modifications of the stem, root or leaf and include rhizomes (underground stems) and storage organs (e.g. bulbs, corms and tubers). Perennial weeds rely on the energy stored in these organs to grow new shoots. Although these vegetative structures usually have a shorter survival time in the soil than seeds, only a very small part is required to grow a new plant.

Characteristics of perennial weeds:

- competitive adaptations
- storage organs
- vegetative reproduction.

Horseweed, an annual weed

Perennial weeds – dandelion and clover

Properties of different weed types	
Annual weeds	**Perennial weeds**
grow for one season only	grow for several seasons
sexual reproduction	sexual and asexual reproduction
rapid growth	vegetative structures (storage organs)
short lifecycle	
high seed output	
long-term viability of seeds	

contd

Pests

The main pests of crop plants are the **invertebrate animals**, especially insects. **Insects** are a threat to food security, since they eat various plant parts and decrease yield. **Nematodes** are very small, non-segmented, colourless roundworms that live in many habitats, but are particularly numerous in the soil. They are plant parasites and attack crop roots. Some attack plants from the outside and others live inside the plant. They may inject saliva into the host plant, killing plant tissue or causing the formation of a gall. They cause stunted growth, chlorosis and distortion of the roots. **Molluscs**, such as snails and slugs, can also damage crops.

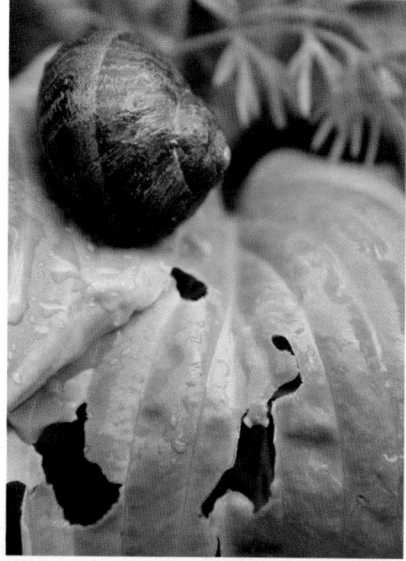

Snail damage to plant

Aphid damage to Russian wheat

Invertebrate	Example	Damage done	Effect
insect	aphid e.g. greenfly	pierces stem and leaves and sucks cell sap	decreased yield; introduction of toxin or virus to plant
	colorado potato beetle	eats leaves	reduced yield
	locust	eats plant parts	reduced yield
nematode worm	potato cyst nematode	feeds on plant roots	reduced yield
mollusc	slugs and snails	eat seedlings, roots tubers, bulbs, bark and flowers	reduced yield

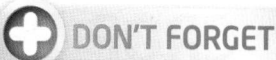

THINGS TO DO AND THINK ABOUT

1 Write a paragraph describing how insect pests are a major threat to food security.

2 Write a few sentences to describe how some named perennial weeds reproduce.

Plant diseases can also be caused by fungi, bacteria and viruses as shown in the table:

Plant disease	Pathogen causing disease	Invertebrate vector
beet mosaic	virus	aphid
cabbage black ringspot	virus	aphid
apple rot	bacteria	flies (maggots)
cabbage blackleg	fungus	maggots

CROP PROTECTION 2

CONTROL OF WEEDS, PESTS AND DISEASES

There are several different methods of controlling crop pests:

1. prevention
2. cultural control
3. chemical control
4. biological control

PREVENTION

Preventing the introduction of weeds to the soil can be easier than removing established weeds. Good practice includes removal of weed seeds and vegetative structures from tools and machinery used to prepare the soil, ensuring seeds are certified to be free of weeds, and controlling weeds in the surrounding areas.

CULTURAL CONTROL

This uses established farming practices to prevent the growth and spread of weeds, while allowing the desired plant or crop to grow quickly so that it can successfully out-compete weeds. Cultural methods of weed control include planting **dense populations** of crop plant, regularly **removing any weeds** which appear in the crop, **composting weeds** or **feeding weeds to livestock** to destroy their seeds.

Fertilisers encourage rapid growth of the desired crop and **mechanical** practices, such as hoeing, pulling out by hand, digging and mulching can also prevent weed growth. **Crop rotation** involves changing the crops which are grown in a particular field each year, preventing the build-up of pests, weeds and diseases which would be detrimental to crop health and vigour. The diagram shows a seven-yearly crop rotation.

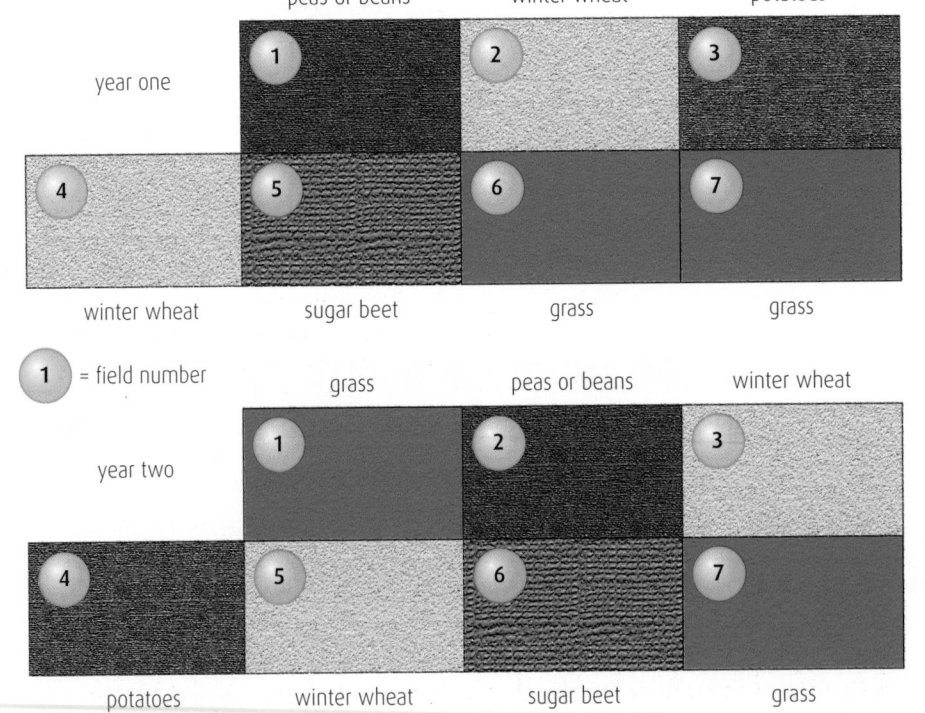

CHEMICAL CONTROL

Selective pesticides

A **selective pesticide** removes only certain types of pests, leaving others unharmed. **Non-selective** (broad spectrum) brands will kill a wide range. Weed control in a field of a particular crop can be accomplished with a selective herbicide which will kill weeds without killing the crop. Lawns can be kept weed free with selective weed killers, which target broadleaved weeds such as dandelions, without killing the grass.

Selective insecticides are available which kill pests such as aphids, without killing their natural predators such as ladybirds. This is beneficial since **biological control** will still be possible (see page 78).

Contact and systemic pesticides

Contact pesticides destroy any plant or animal tissues they come in contact with and so do not destroy the underground vegetative structures of perennial weeds. Household fly spray is an example of a contact insecticide.

Systemic pesticides move from the site of application to different untreated tissues within the animal or plant. They can be applied by spraying the plant or animal or by drenching the soil with the chemical, which is then absorbed into the organism's tissues.

A systemic insecticide or fungicide will move within the plant via the xylem and/or phloem to roots, leaves or stems. An insecticide kills insects such as greenfly when they suck plant sap from the phloem tissue and a fungicide will give the plant protection from fungal attack, usually with a more prolonged effect than a contact fungicide. Some systemic insecticides can be applied to an animal and move through its body to control pests such as lice or fleas.

Systemic herbicides are absorbed and transported to all parts of a weed, leading to its death. They are effective at controlling perennial weeds by preventing regrowth from vegetative structures and, although slower acting, are more effective.

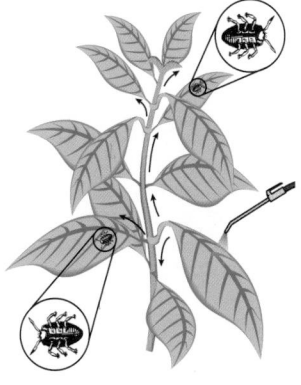

Systemic pesticides are transported from their application site to other plant tissues

PROBLEMS WITH PLANT PROTECTION CHEMICALS

Chemicals that are used to protect crops and other food plants from damage can cause problems. Some of these problems are outlined in the table.

Problem	Explanation
Toxicity	Some pesticides are poisonous to animals, including humans, and can cause diseases such as cancers.
Persistence	A persistent pesticide is not broken down in the environment (it is not biodegradable). Therefore, if it is toxic, its harmful effects will be prolonged.
Accumulation (bioaccumulation)	Persistent pesticides pass along food chains and build up (accumulate) in animal tissues (e.g. mercury in fish). Their concentration will be higher than that in the environment.
Magnification (biomagnification)	The concentration of persistent pesticides increases at each trophic level and will harm the larger predators at the top of the food web (e.g. DDT insecticides in birds of prey).
Resistant populations	A few organisms within a species may not be killed by the pesticide and are said to be resistant to it. Natural selection favours the resistant individuals. When they reproduce, they pass on the genes for this resistance. If the pesticide continues to be used, **selection pressure** results in an increase in the number of resistant organisms in each generation until the entire population is resistant.

DON'T FORGET

An **adaptation** is an inherited characteristic that makes an organism well suited to survival in its environment or niche. Organisms which are better adapted to their environment are more likely to survive, reproduce and pass on their favourable genes to their offspring. This is **survival of the fittest** and is the basis of **natural selection**.

ONLINE

The article 'Crop pests: under attack' (available at www.brightredbooks.net) describes other threats to food security, including some effects of climate change.

ONLINE TEST

Once you've learned about this topic, test your knowledge at www.brightredbooks.net

THINGS TO DO AND THINK ABOUT

1   Read the article on Crop pests. Summarise the main problems and the possible remedies.

Potato cyst nematodes live on roots of potato plants and, at high densities, damage the roots and cause early death. Crop rotation, with at least six years between plantings of susceptible crops, can reduce the number of nematodes. Investigate the control and management of this disease using the links at www.brightredbooks.net

CROP PROTECTION 3

DISEASE FORECASTS

Plant disease forecasting is used to predict the likelihood of the occurrence of a plant disease. Fungal diseases, in particular, are influenced by climate and weather patterns and so collaboration between plant pathologists and meteorologists allows predictions about the risk of disease. This enables farmers, for example, to make decisions about applying a particular fungicide to protect crops before they become diseased, which is more effective than treating a diseased crop.

DON'T FORGET

Biological control is a method of reducing numbers of pests using their natural enemies, such as predators, parasites or pathogens. A parasite is an organism which lives in or on the body of another organism, its host. The parasite benefits and the host is harmed.

BIOLOGICAL CONTROL

In their natural habitat, insects are controlled by other organisms and the environment; this is natural control. In **biological control**, human intervention introduces the natural enemy of the pest (referred to as the **control agent**). If biological control is used successfully, instead of chemicals, potentially harmful effects of pesticides are avoided. Biological control can be used to control weeds, plant diseases and insect pests.

Example: 1 Control of glasshouse whitefly

Glasshouse whitefly is a common pest in greenhouses and has a large range of host plants including tomatoes, cucumber, peppers, grapes and beans. The flies suck sap from the phloem, which results in reduced plant vigour, stunted growth and leaf fall. They also excrete a sticky substance, honeydew, on the host plants and this allows mould to develop on the leaves.

Insect populations reproduce rapidly and many pesticide-resistant strains have developed; therefore, **biological control** tends to be more effective than insecticides. The tiny parasitic wasp *Encarsia* attacks whitefly nymphs and was one of the first agents of biological control. All stages of its lifecycle occur within its host; females lay their eggs in the immature stages (scales) of the whitefly. Ten to 14 days later the scales turn black and the immature whitefly are destroyed. The scales are also eaten by the adult wasps. After around another two weeks, the adult wasp emerges by cutting the blackened scale.

Example: 2 Control of the red spider mite

The glasshouse red spider mite is another destructive greenhouse pest. They are extremely small, so it is easier to look for the damage they have caused. They pierce leaf cells and suck plant juices, resulting in discolouration of the leaves, stunted growth and death of the plant. A natural predator of the red spider mite, the predatory mite *Phytoseiulus*, is the main **biological control** agent. **Chemical control** of red spider mites using insecticides can also be effective. However, these insecticides also kill the mites' natural predators which can aggravate the problem.

A red spider mite (only red at certain stages of its life), magnified a million times

ONLINE

Read more about red spider mites and their control by following the link at www.brightredbooks.net

contd

Example: 3 Control of butterfly caterpillars

The bacterium *Bacillus thuringiensis* occurs naturally in the soil and on plants, producing a crystal protein that is toxic to specific groups of insects including butterfly caterpillars.

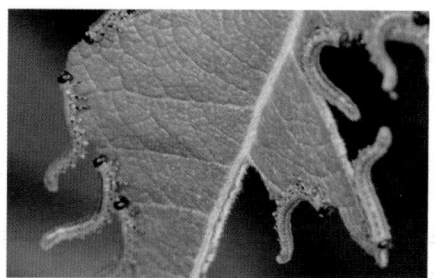

Risks with biological control

There are, however, some risks associated with biological control:

- non target insects may be harmed
- crops may be damaged by the control species i.e. the control species becomes a pest.

Example: 1 Introduction of cane toads to Australia

Cane toads were introduced to Australia as a biological control against cane beetles which were destroying sugar cane crops. However, the beetles normally feed on top of the sugar cane which can grow up to 8 metres in height and since cane toads cannot climb they couldn't reach the beetles. Also the beetles usually feed during the day whereas the cane toads feed at night. The sugar cane fields in Australia are much drier than that of the toads' native habitat and so the toads moved quickly southward through Australia to moister areas and have spread in numbers rapidly. They feed on many native species and out-compete others for food. They are prolific breeders and are poisonous if eaten by predators and are now considered an invasive species and a pest.

Example: 2 Introduction of Harlequin Ladybirds to North America and Europe

Harlequin ladybirds were introduced to North America and mainland Europe as a biological control for aphids. However, Harlequin ladybirds are larger than many other aphid predators and eat larval stages of other ladybirds and by killing aphids they leave less food for other ladybird species. The decline in native ladybird species in the UK and USA has been directly attributed to them. When aphids are scarce, they may feed on non pest species such as butterflies. Also, Harlequin ladybirds are a pest of apples and pears and from late summer feed on soft fruit causing blemishes on the fruit.

Integrated pest management

Integrated pest management (IPM) is a process for controlling pests while minimising any risks to other organisms, including humans, and the environment. It involves a range of techniques, using pesticide as a last resort.

The techniques and processes include:

- **identification** of the pest(s)
- **monitoring** numbers of pest(s) and damage caused
- long-term **pest prevention** by controlling environmental factors to create unfavourable conditions for the pest
- using a **combination of** control methods which may be more effective than a single method of control.

Many IPM programmes combine cultural, chemical and biological control.

ONLINE

Find out more about IPM and biological control at www.brightredbooks.net

ONLINE TEST

Head online and test yourself on this at www.brightredbooks.net

THINGS TO DO AND THINK ABOUT

1 When a caterpillar eats the crystal protein produced by *Bacillus thuringiensis* and it reaches the gut, the crystals dissolve in the alkaline gut contents and proteins are released. These proteins damage the gut lining; the caterpillar stops feeding and starves. Why is it that the bacterium is harmless to humans and other animals?

2 Describe risks associated with biological control using the 'Biological pest controls' link at www.brightredbooks.net/

ANIMAL WELFARE

The UK has a huge sheep farming industry with an estimated 32 million sheep and lambs in 2012. At any one time, it has 5 million pigs on its farms and millions of chickens are produced in the UK every week. In 2014, the UK dairy herd had risen to almost 1.9 million and, although it is decreasing, the beef herd was estimated to be around 1.5 million. Increasing meat production helps to ensure food security, but there are concerns for animal welfare. Animal welfare is the wellbeing of animals and many different approaches have attempted to define good animal welfare.

Traditionally, an animal's welfare was based on whether it was producing well, e.g. growing quickly, producing lots of milk or eggs and so on. However, some highly productive farm animals do not experience positive welfare, for example battery hens.

DON'T FORGET

Intensive farming is necessary to meet the food demands of the ever increasing human population.

ONLINE

Watch the introduction to animal welfare at www.brightredbooks.net

DON'T FORGET

Intensive 'battery farming' is a practice whereby animals are reared in a restricted space, usually indoors.

ONLINE

Find out about the five freedoms for animal welfare at www.brightredbooks.net/

ONLINE

Read about some stereotypies at www.brightredbooks.net

THE FIVE FREEDOMS FOR ANIMAL WELFARE

A list of recommended conditions, which should be applied to animals in order to maintain their wellbeing, was originally developed by the UK's Farm Animal Welfare Council (FAWC) and is now internationally recognised. The 'Five Freedoms' are based on three principles; living a natural life, being fit and healthy, and being happy. They include:

- freedom from thirst and hunger
- freedom from discomfort
- freedom from pain, injury and disease
- freedom to express normal behaviour
- freedom from fear and distress

BEHAVIOURAL INDICATORS

Behavioural indicators are measures of animal welfare. These are summarised in the table.

Behavioural indicators of good welfare	Behavioural indicators of poor welfare
alertness; curiosity; engages in play; animal performs a range of activities; animal interacts with group and with humans	limited activity; no response to stimuli or play; misdirected behaviour such as abnormal fear or aggression towards humans; fighting; failure in sexual behaviour; failure in parental behaviour; sickness or pain behaviours; *stereotypies, for example caged sows bite bars, calves exhibit eye rolling and caged chickens may peck at their feathers

*A **stereotypy** is a repetitive behaviour which serves no obvious purpose and is constant in form.

Sows in cages without straw or dirt may bite the cage bars until their gums bleed

COSTS AND BENEFITS OF DIFFERENT LEVELS OF ANIMAL WELFARE IN LIVESTOCK PRODUCTION

Good animal welfare		
Costs	**Benefits**	**Ethical impact**
Increased costs to the farmer of free range farming methods due to requirement for more straw, food, space and labour	Higher yield and quality of products Greater breeding success and effective parenting Improved nutritional value of food products	Better quality of life for animals Less stress due to overcrowding
Poor animal welfare		
Costs	**Benefits**	**Ethical impact**
Lower yield and quality of products, such as meat, eggs, milk and wool Reduced breeding success and rejection of young Lowered nutritional values of products Animals more susceptible to disease	Cheaper products	Poorer quality of life for animals Limited opportunities to display natural behaviour and increased likelihood of abnormal behaviour, such as tail-biting, cannibalism, and feather pecking Farmers may carry out invasive procedures such as as castration, ear notching or beak trimming

MEASURING MOTIVATION

Anthropomorphism is the attribution of human characteristics and qualities to non-human beings, including animals. People have often debated whether animals have feelings and emotions; feelings are subjective and cannot be investigated directly. However, it is possible to determine how animals feel about particular conditions by offering preference and motivational tests.

A preference test is an experiment where an animal has the choice of conditions and shows which condition the animal prefers. Scientists can use a T maze or a Y maze with different options available at the ends of the maze. Animals learn which options are available and their choices are then counted over a number of tests.

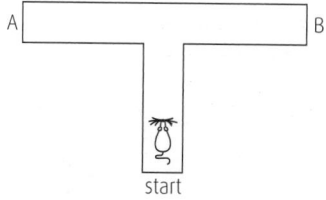

ONLINE

Read about assessing animal welfare at www.brightredbooks.net

Simple T-shaped maze

SCIENTIFIC STUDIES OF ANIMAL BEHAVIOUR

Ethology

Ethology is the study of animal behaviour, particularly that which occurs in the animal's natural or semi-natural environment. For example, the natural habitat of the pig is woodland and an equivalent semi-natural environment would be an enclosed area of woodland. However, on intensive farms pigs are kept inside in sheds or pens which may not be straw-lined.

An **ethogram** is a list of typical species behaviours, including individual, pair and group behaviour. Information on the frequency and duration of the behaviours is collected by observation (sometimes with the use of aids such as camera or video recorder). When an ethogram is analysed, hypotheses can be made and these can be tested by investigation. This allows the most favourable environments for different species of domesticated animals to be identified.

ONLINE

Read about preference and motivational tests at www.brightredbooks.net

ONLINE TEST

How well have you learned this topic? Take the test at www.brightredbooks.net

THINGS TO DO AND THINK ABOUT

1 Research stereotypy and make a table giving examples of stereotypic behaviour in several different animals. Include the name of the animal and a description of the behaviour.

2 Write a short paragraph to describe how a T-maze or a Y-maze could be used in a preference test.

SYMBIOSIS

DON'T FORGET

Other living things are an important feature of an organism's ecosystem because organisms depend on one another, for example for food or for a habitat.

ONLINE

Read about different forms of symbiosis at www.brightredbooks.net

ONLINE

Find out more about symbiosis and co-evolution at www.brightredbooks.net/

AN INTRODUCTION TO SYMBIOSIS

Some organisms depend on other organisms because they form a partnership and, so, **co-evolution** occurs. This means they evolve together over millions of years and the evolutionary change of one organism is triggered by its interaction with the other. Symbiotic relationships are formed through co-evolution. Symbiosis means 'living together' and refers to an intimate relationship between organisms from two different species

Different types of symbiosis exist, including **parasitism** and **mutualism**.

PARASITIC RELATIONSHIPS AND TRANSMISSION

A parasite lives in or on the body of another organism, which is known as the **host**, and gains energy or nutrients from it. Therefore, the parasite benefits from the relationship and the host is harmed, due to a loss of valuable resources. Parasites such as ticks, fleas, leeches, and lice live on the surface of the host's body, whereas some live inside the host, including tapeworms, roundworms and flukes, as well as bacteria and viruses.

The majority of parasites cannot survive in the absence of a host. Due to their limited metabolism, they need to live in or on a host organism. Since they are adapted to live in or on a host, they may have lost some of the organs which are necessary to survive on their own.

Parasites usually have a limited metabolism:
- Tapeworms attach themselves to the insides of the intestines of animals and obtain partially digested food from the host.
- Fleas bite the skin of their host and gain energy by sucking their blood. Many types of parasites carry and transmit disease in this way.

It is important to the parasite that it can be transmitted from one host to another. This can happen by direct contact, via resistant stages or via a vector.

Direct contact

This is when a parasite is passed directly from one host to another when the two hosts come into close physical contact with one another; head lice can only move from one human head to another by direct contact.

Resistant stages

Some parasites, for example protozoa, have resistant or resting stages (cysts or oocysts) which can resist drying out and other stresses such as temperature extremes and harsh chemicals. They can survive out with the host for a period of time.

Electron micrograph of human head louse

Vectors

A vector is an intermediate organism that transfers a parasite from one host to the next. An example of a vector is a species of mosquito which transfers the protozoan parasite *Plasmodium*, which causes malaria, from one human host to another

ONLINE

Read about the lifecycle of the malaria parasite using the link at www.brightredbooks.net

Secondary host

A secondary or intermediate host is necessary for some parasites to complete their lifecycle. Asexual reproduction or the development of larval stages of the parasite may occur within the secondary host. Humans are the secondary host for malarial parasites.

1. Victim is bitten by infected mosquito. Mosquito injects protozoan into human.

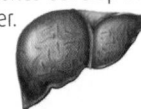
2. Parasites develop in the liver.

3. Parasites finish maturation in the red blood cells and release gametes.

4. Mosquito takes blood – including the parasite gametes.

5. Fertilization and development of zygote happens within the mosquito vector.

MUTUALISM

Mutualism is a form of symbiosis which benefits both species involved.

Example: 1 Herbivores

Herbivores must get their energy from the plant material which they eat. However, their digestive systems do not produce cellulase, the enzyme necessary to digest the cellulose in plant cell walls. They need the help of cellulose digesting bacteria and protozoa which live in their digestive system. In return, these microorganisms get food from the gut of the herbivore.

Example: 2 Coral Polyps

Coral polyps are animals and they form a mutualistic relationship with zooxanthellae algae which live within the polyp cells. Coral provides the algae with carbon dioxide for photosynthesis and other wastes including water and ammonium compounds. The algae provide the coral with sugar and oxygen.

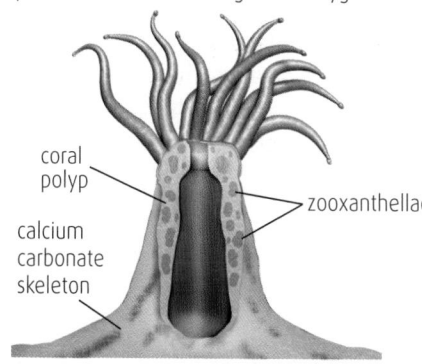

Diagram of zooxanthellae in coral polyp

coral polyp
zooxanthellae
calcium carbonate skeleton

 ONLINE

Watch an example of mutualism at www.brightredbooks.net

Mutualism between cleaner wrasse and larger fish

EVOLUTION OF MITOCHONDRIA AND CHLOROPLASTS

It is thought that mitochondria and chloroplasts in eukaryotes evolved from free living bacteria (prokaryotes) which were engulfed by other cells, forming a mutualistic relationship. The prokaryotic cells would benefit from protection from the larger cell and the eukaryote would gain:

- more energy if it took in an aerobic prokaryote
- a food supply from photosynthesis if it engulfed a photosynthetic prokaryote.

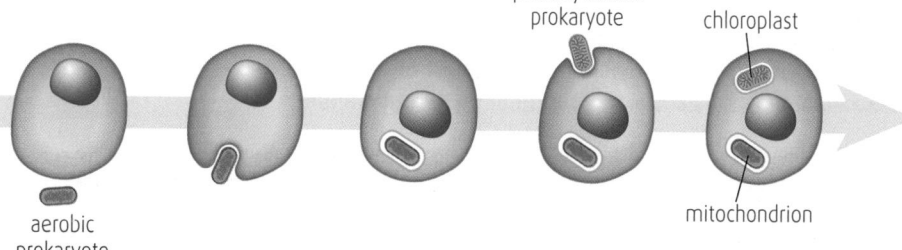

photosynthetic prokaryote
chloroplast
mitochondrion
aerobic prokaryote

DON'T FORGET

Prokaryotes do not have membrane-bound organelles such as a nucleus, chloroplasts or mitochondria. Plant, animal and yeast cells are all eukaryotes and have a membrane-bound nucleus and other membrane-bound organelles, such as mitochondria or chloroplasts.

 ONLINE

Watch the animation showing the evolution of these organelles at www.brightredbooks.net

 ONLINE TEST

Once you've learned about this topic, test your knowledge at www.brightredbooks.net

THINGS TO DO AND THINK ABOUT

1 Describe how a parasite is transmitted to a new host by:

   a direct contact
   b resistant stages
   c a vector
   d a secondary host

2 Explain the differences between parasitism and mutualism.

SOCIAL BEHAVIOUR 1

SOCIAL HIERARCHY

Social hierarchy in wolves

Social hierarchy (also known as dominance hierarchy) is present in many animal groups. Hierarchies are usually established by fighting or display behaviour which results in a ranking of the animals from the most dominant to the most subordinate. Each animal is dominant over those below it in rank and submissive to those above it, determining access to food and a mate. Dominance hierarchies are common in social mammals, such as wolves, and in birds.

The term 'pecking order' comes from dominance hierarchy in chickens where the most dominant can peck any other chicken and the lowest ranking chicken is pecked by all the others. A dominance hierarchy reduces fighting and aggression, conserves energy and prevents injury. The more dominant members will mate and pass on their genes to the next generation.

COOPERATIVE HUNTING

Cooperative hunting is when a group of animals work together to gain food; this occurs in several species, including lions, wild dogs, hyenas, chimpanzees and some birds of prey and large fish. Hunting cooperatively gives more opportunities to attack and kill a prey animal before a herd of prey scatters and escapes. Sharing of food with the whole group will occur if the energy gained by sharing will exceed that obtained by individual foraging. Cooperative hunting benefits both dominant and subordinate animals. Advantages include:

- an increase in hunting success rate
- predators can take down larger prey than they could do individually
- the kill can be made more efficiently
- all predators obtain the maximum amount of food possible
- subordinates may gain more food than by foraging alone.

Remember, however, that not all animals which hunt in groups do so cooperatively. They may be together because that is where the food is, for example many gannets hunting shoals of fish are working individually, not cooperatively.

Lions working cooperatively to take down a large buffalo

DEFENCE

Living in a social group can also protect animals.

Behaviour	Benefit	Example
group organisation ensures a lookout is always available to make alarm calls	warns group of danger from predator	meerkats birds
group works together to mob or attack a predator	protects offspring	chimpanzees crows and gulls
animals move together in a herd, flock or school	it is more difficult for predators to pick off individuals from a group	zebras birds fish
travelling in a group	provides protection for most of herd	elephants

ALTRUISM

Altruistic behaviour is common in animals and is a behaviour which benefits the recipient but harms the donor. It involves an animal acting in a way that will decrease its own survival chances, but improves the survival chances of another. For example, vampire bats regurgitate blood and give it to other members of their group who have failed to feed. Vervet monkeys and ground squirrels give alarm calls to warn other monkeys of the presence of predators, even though they may attract attention to themselves, increasing the chance of being attacked. In social insect colonies (ants, wasps, bees and termites), sterile workers care for the queen, build and protect the nest, forage for food and care for the larvae.

KIN SELECTION

Altruism is common between a donor and a recipient if they are related (kin). Since close relatives share many alleles, natural selection will favour behaviour that increases the survival of relatives (kin selection). The donor benefits in terms of the increased chance of survival of shared genes in the recipient's offspring or future offspring. For example, vampire bats are more likely to share a blood meal with kin.

Helper behaviour

In birds, a breeding pair may get help in raising young from 'helper' birds, usually their older offspring, who will protect the nest from predators and help to feed the young.

Reciprocal altruism

This is a type of altruism in which the roles of the donor and recipient later reverse; an organism temporarily reduces its own fitness while increasing the fitness of another organism, with the expectation that the other organism will act in a similar manner at a later time. It occurs in social animals and social insects. The prisoner's dilemma is a simple model of reciprocal altruism.

Relatedness

Relatedness is the probability that two individuals share an allele due to recent common ancestry and is expressed as the *coefficient of relatedness* (r). It ranges from 0 (unrelated) to 1 (clones or identical twins). Children inherit 50% of their alleles from each parent; so, $r = 0.5$ for parents and children.

ONLINE

Read about the reciprocal altruism and prisoner's dilemma at www.brightredbooks.net

ONLINE

Read about altruism and kin selection at www.brightredbooks.net

 THINGS TO DO AND THINK ABOUT

1   Give definitions of:
    - social hierarchy
    - altruism
    - reciprocal altruism
    - kin selection

2   Describe the benefits of cooperative hunting

ONLINE TEST

Head online and test yourself on this at www.brightredbooks.net

SOCIAL BEHAVIOUR 2

Ant queen beginning to dig a new colony

ONLINE

Watch a honey bee waggle dance at www.brightredbooks.net

ONLINE

Watch social behaviour in ants at www.brightredbooks.net/

ONLINE

Read about social behaviour of termites at www.brightredbooks.net

ONLINE

Read about the importance of pollinators to food security at www.brightredbooks.net

SOCIAL INSECTS

Many insects exhibit social behaviours which means that they live together in large, organised family groups and exhibit a range of complex behaviours. A truly social insect society will have:

1 cooperative care of the young insects
2 parents and their offspring living together
3 the development of a caste system.

Ants, bees, termites and wasps live in colonies where different types of adults play different roles. Reproduction and survival depend on all members of the colony working together. The majority are workers who labour together to raise their relatives. The success and efficiency of the group depends on the workers foregoing reproduction, which allows them to focus all their efforts on their other tasks including defence of the colony. This ensures their genes are passed on to the next and following generations through their breeding relatives. Although, only some individuals in the colony contribute reproductively, colonies produce large numbers of offspring.

The social structures of insect colonies often rely on effective communication. For example:

1 Honey bee workers perform a waggle dance to teach the other workers the direction and the distance to a particular food source.
2 Some insects use chemical messengers known as pheromones. Certain ants, after finding food, lay down a pheromone trail when returning to the nest. This attracts and guides other ants to the food source.

Examples of social behaviour in insects

Insect	Adult types	Number in colony	Role/tasks carried out	Benefits
honey bee	queen	one	reproduction and egg production	since the queen is the only female in the colony which reproduces, all the workers are her daughters; therefore, the workers share many of the queen's genes which will pass to future generations
	workers (all females)	several thousand	build nest, collect food, rear the brood	
	drones (males)	several hundred	impregnation of queen (fertilise her eggs)	
ant	queen	one (usually)	reproduction and egg production	
	workers (sterile females)	many	forage for food, maintain nest, tend to brood	
	soldiers (sterile females)	many	defend colony	
	drones (males)	many	mate with queen	

ECOLOGICAL AND ECONOMIC IMPORTANCE OF SOCIAL INSECTS

Keystone species are those which have a particularly important impact on their ecosystem. Without keystone species, the species diversity of the ecosystem would decrease and the ecosystem may cease to exist. Social insects are often keystone species. Those which benefit humans and are of economic importance are said to provide **ecosystem services**. These include the pollinators, decomposers and those that are used in the control of pests, for example parasitic wasps.

It has been estimated that over a third of global food production is from crops that depend on animal pollinators so if their numbers decrease then food security will be jeopardised.

contd

Keystone species	Role
honey bees and wasps	pollinate flowering plants, including many food crops
termites	build mounds which provide habitats for other species; provide protein-rich food for many animals; are decomposers, so recycle soil nutrients
ants	pollination, nutrient turnover, seed dispersal

DON'T FORGET

Pollination is the transfer of pollen from anther to stigma in a flowering plant. Without pollination plants could not reproduce.

PRIMATE BEHAVIOUR

Humans are **primates**, a large and diverse group of mammals including monkeys and apes. The long period of parental care in primates gives time for young to learn complex social behaviours by watching and copying others. Chimpanzees learn to make tools such as termite fishing sticks and leaf sponges. Termites are a popular food for chimps and they have been observed stripping leaves off twigs, pushing them into termite mounds and after a short time pulling them out covered in termites. They have also been seen chewing leaves and using them to absorb water from puddles which they then drink.

Most primates live in social groups where there is competition for resources (such as food or a mate), so conflict can arise. Living in a group, however, does provide protection from predators and of food sources. Primate species exhibit a range of complex behaviours which help to support the social structure of the group and reduce unnecessary conflict. The social structure of each species depends on:

- its particular ecological niche – the role it plays in its ecosystem
- the distribution of its resources – the type of food eaten and its availability
- the taxonomic group to which it belongs – different species display different behaviours.

Dominance hierarchies are common in primates and help to keep order, avoiding the heavy costs (injury and energy) associated with fighting. Social interactions between group members can be competitive (**agonistic**) or cooperative and may include facial expressions, body posture, vocalisations, sexual presentation and other behaviours.

Alliances

In some primate social groups, individuals may become **allies** in order to maintain or improve their social status. In chimpanzees, alliances allow males to dominate another individual which they would be unable to do alone and may enable allies to challenge a higher ranking male. The more alliances a male chimpanzee can form, the more chance he has of being dominant.

Ritualistic behaviour

A **ritualistic display** is a sequence of repeated behaviours used to communicate in reproductive or agonistic situations. A dominant animal may signal aggression with an open mouth and a particular vocalisation. For example, an alpha male Western gorilla makes hooting sounds, pounds his chest, kicks his legs and runs sideways if approached by another male, in order to intimidate him. However, there is no physical contact. Some use signals to indicate their subordination to the dominant animal and, in many primates, subordination is signalled with a gesture that looks like a smile.

Appeasement behaviours

Tense situations can be resolved through **appeasement behaviour** in which one animal tries to reduce the aggression of another. Grooming, hugging and kissing often follow an agonistic interaction between primates to reduce tension and repair the relationship.

ONLINE

Read about different social structures in primates at www.brightredbooks.net

ONLINE

Read about alliances in chimpanzees at www.brightredbooks.net

ONLINE

Read about facial expressions in chimpanzees at www.brightredbooks.net/

ONLINE TEST

How well have you learned this topic? Take the test at www.brightredbooks.net

THINGS TO DO AND THINK ABOUT

1. Explain what is meant by keystone species and ecosystem services.
2. What is appeasement behaviour?
   Give an example of appeasement behaviour and explain why a named primate may use this type of behaviour.
3. How might an increase in social status benefit a male chimpanzee?

BIODIVERSITY

DON'T FORGET

Biodiversity is the variety of living things on Earth.

ONLINE

Read about the importance of these databases and the associated difficulties at www.brightredbooks.net

CENTRAL DATABASES OF SPECIES

Current estimates suggest there may be over 9 million different species of living things on Earth and there are about 2 million named species. About 50% of the known species are animals and most of the animals are insects. Most of the known vertebrates are fish and there are about 0.25 million known species of flowering plants. There are several existing databases which attempt to list every known species.

MASS EXTINCTION

The major mass extinctions

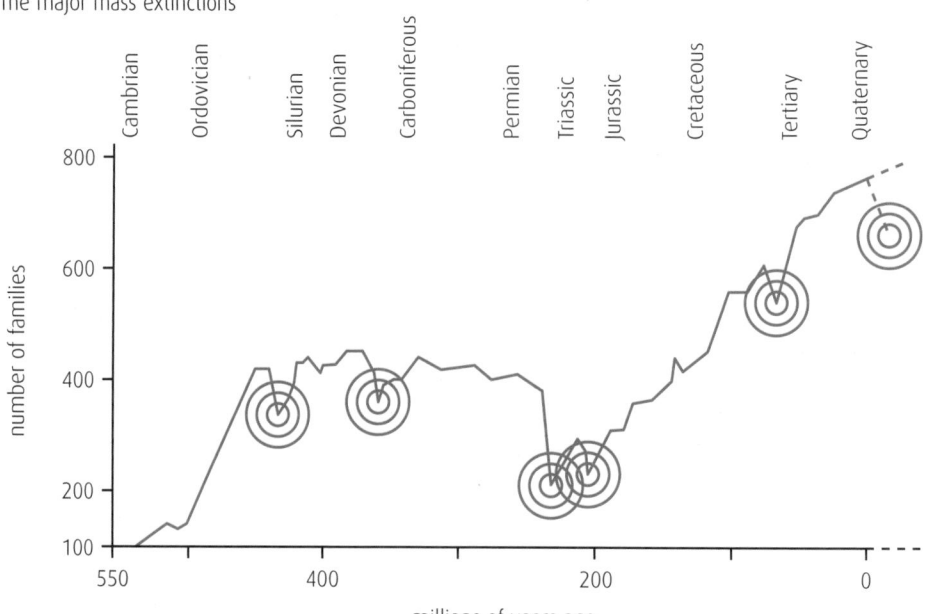

ONLINE

Read about mass extinction events at www.brightredbooks.net

ONLINE

Read about adaptive radiation of mammals at www.brightredbooks.net

ONLINE

Read about the difficulties in estimating extinction rates at www.brightredbooks.net

A **mass extinction** is a rapid decrease in numbers of species on Earth; 25–70% of all species may cease to exist over a period of tens of thousands of years, as for example in the extinction of the dinosaurs. When organisms die their remains may be preserved as fossils, providing a record of their existence. Fossil evidence shows that several mass extinction events have occurred in the past and that the Holocene extinction which began 11 500 years ago is still ongoing.

Increasing biodiversity after a mass extinction

A mass extinction leaves many unfilled niches and the surviving organisms may take advantage of these. New species evolve from a common ancestor to fill these empty niches, as happened when mammals diversified into the empty niches occupied by dinosaurs. This is **speciation**. Therefore, following a mass extinction is a period of recovery as biodiversity is slowly regained. Fossil evidence indicates this takes around 10 million years.

Estimating extinction rates

The background extinction rate is a measure of how often extinctions happen naturally without the interventions of humans. Most estimates of the rate of extinction of a species are derived from habitat surveys – however, this is extremely difficult to estimate accurately.

contd

The extinction of mega fauna

Mega fauna were species of very large animals, usually weighing well in excess of 44 kg. Examples include the woolly mammoth (from around 6000 kg) and giant bison of North America (up to 2000 kg). There were a dramatic number of extinctions of mega fauna species during the Pleistocene era, which coincided with the spread of humans into several continents. Climate changes, as well as hunting by humans, are likely to have contributed to these extinctions.

The role of humans in extinction of species

The current rate of species extinction is much greater than the background rate. This is due to the rapidly increasing rate of ecosystem degradation by humans, which endangers the survival of many species. The main causes of ecosystem degradation by humans are:

- deforestation – cutting down forests to make way for agriculture
- pollution of air, water and land
- over exploitation of resources, for example overfishing and overhunting
- destruction of habitats, for example by the removal of hedgerows.

mega fauna

deforestation

MEASURING BIODIVERSITY

There are three components of biodiversity which can be measured: genetic diversity, species diversity and ecosystem diversity. This allows scientists to compare the biodiversity of different areas.

Genetic diversity

The **number** and **frequency** of alleles in a population can be used as a measure of **genetic diversity**. Genetic diversity is the total genetic variability or the total number of genetic characteristics of a species. If a particular population dies out, then alleles are lost and genetic diversity decreases. Lack of genetic diversity is an issue because, if environmental conditions change, the species may not be able to adapt. For example, a lack of genetic diversity led to the Irish potato famine since only two varieties of potatoes were planted and neither was resistant to the fungus responsible for the blight.

Species diversity

This is a measure of the number of different species in an ecosystem (the **species richness**) *and* the proportion of each species (its **relative abundance**). These measures can be used to calculate a biodiversity index of an area.

Example:

Two areas, X and Y, have the same species richness: each has five different species.
Area X has 100 organisms living in the area with 20 of each species. Area Y also has 100 organisms but has 80 of one species and only five of the other four species.
Therefore **area X has greater species diversity** since it is not dominated by one species as in area Y.

Ecosystem diversity

Ecosystem diversity is the number of distinct ecosystems within a defined area. A region with a wide variety of ecosystems will have greater species diversity than an area with only a few. Biogeography is the study of the distribution of species and ecosystems, and island biogeography investigates the factors that affect the species richness of islands. In this context, an island is any isolated natural community and, as well as a traditional island (a piece of land surrounded entirely by water), can also refer to mountain peaks and desert springs, for example. The more remote and isolated a **habitat island** is, the lower its species diversity. Species diversity is also affected by the size of an island – islands with a smaller area will have lower diversity.

DON'T FORGET

High species diversity results in a stable ecosystem with complex food webs, while low species diversity gives an unstable ecosystem with simple food webs; one small change could cause many species to die.

Species	Abundance	
	Area X	Area Y
1	20	80
2	20	5
3	20	5
4	20	5
5	20	5
Total no. of organisms	100	100

ONLINE TEST

Once you've learned about this topic, test your knowledge at www.brightredbooks.net

THINGS TO DO AND THINK ABOUT

1. Explain why there is an increase in biodiversity following a mass extinction event.
2. Give the two components of genetic diversity.
3. Explain why low genetic diversity increases the risk of a species becoming extinct.

THREATS TO BIODIVERSITY 1

DON'T FORGET

Ecosystem degradation endangers the survival of many species.

ONLINE

Read about the consequences of overexploitation of fish and whales at www.brightredbooks.net

ONLINE

Gel electrophoresis is a method for measuring genetic variation in natural populations. Read about the use of this technique to monitor harvest species at www.brightredbooks.net

EXPLOITATION AND RECOVERY OF POPULATIONS

Exploitation is the harvesting of a natural resource, for example forests for timber and animals such as fish for food. **Overexploitation** has been responsible for the decline of several natural resources, since they are harvested at a greater rate than they can be replaced. If a particular population becomes extinct, then alleles are lost and genetic diversity decreases. However, if populations do not become extinct, a reduction in overexploitation can lead to a recovery of the population and genetic diversity will increase again.

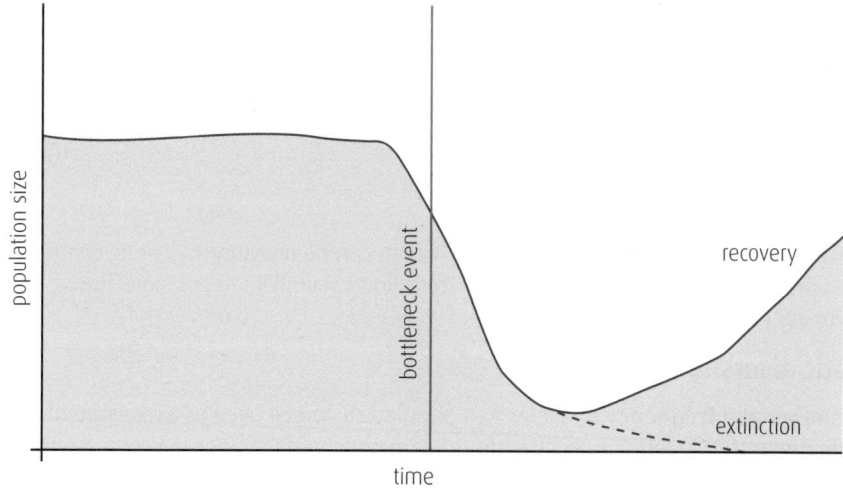

Loss of genetic diversity can be critical for many species. The remaining members of the population will be very similar therefore inbreeding and poor rates of reproduction result. Small populations with reduced genetic diversity are less capable of adapting to a changing environment. This can lead to the phenomenon known as the **bottleneck effect**.

The bottleneck effect

A **population bottleneck** occurs when a population reduces sharply in number, as a result of a fire, drought, overhunting or a disease epidemic, for example. The remaining organisms within the population may be descendants of only a small number of individuals and, therefore, lack genetic variation. Alleles which were rare in the original population could be more common in the new population. The lack of variation means the whole population will be susceptible to any change in environmental conditions; a particular disease could affect the whole population since there would be no resistant strains. Around 10 000 years ago, there was a genetic bottleneck which caused the extinction of several cheetah species, leaving only one. Therefore, cheetahs are now an endangered species and are at risk of extinction.

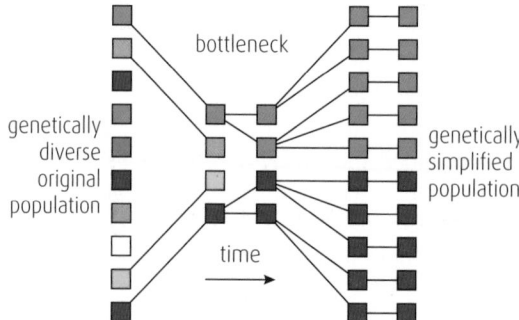

ONLINE

Find out about low genetic diversity in cheetahs at www.brightredbooks.net

There are some species which naturally have a low genetic diversity in their population and yet remain viable, for example certain species of albatross.

HABITAT LOSS, HABITAT FRAGMENTS AND THEIR IMPACT ON SPECIES RICHNESS

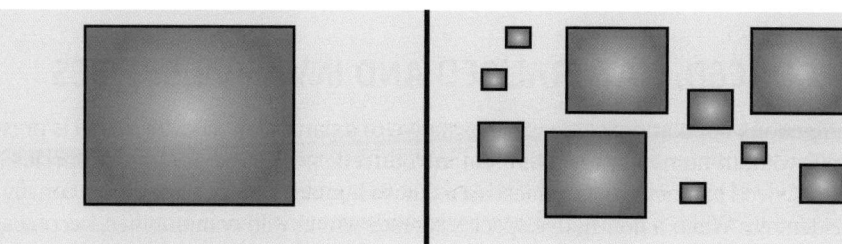

Fragmented landscape versus block of habitat

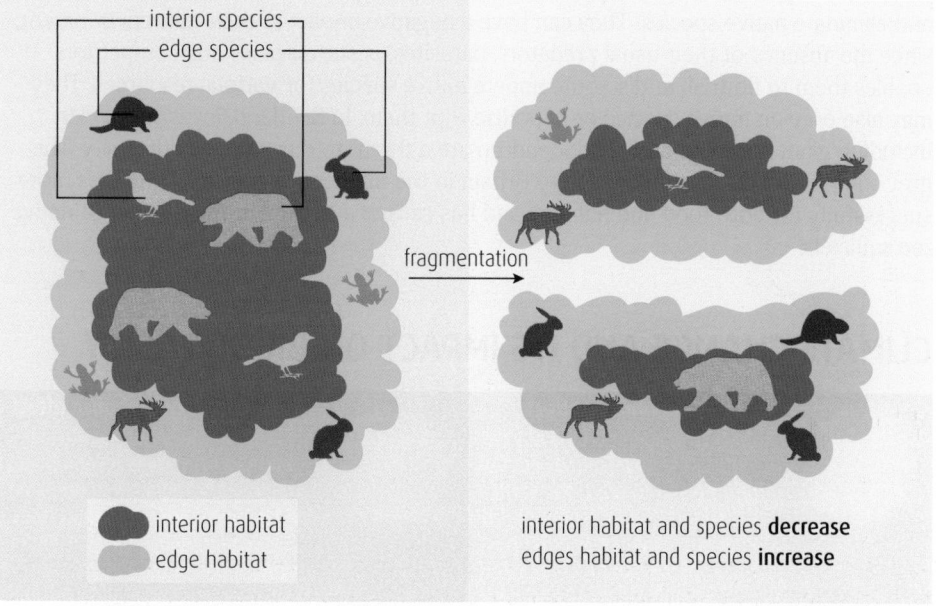

interior species
edge species

fragmentation

interior habitat
edge habitat

interior habitat and species **decrease**
edges habitat and species **increase**

 **DON'T FORGET**

A major cause of modern species extinction is the loss of habitats, often as a result of the actions of humans.

Habitat fragmentation

This occurs when a habitat is divided into several smaller habitats, by road building, agriculture or urbanisation, for example. In Britain, prior to human activity, there were vast areas of native forest habitats, whereas now forest occurs in a much reduced area in smaller fragments. Fragmentation causes a decrease in species richness (reduced biodiversity). Fragments tend to suffer degradation at their edges, which can further reduce their size. These edges contain a small number of species which are adapted to these areas but, as a result of degradation, the **edge species** may move into the interior of the fragment, creating competition with species already there so further reducing biodiversity.

Habitat corridors

Movement between fragments can be dangerous; if animals have to cross a road for example. Fragments can be linked with **habitat corridors** to make movement safer. These are thin strips of habitat linking larger patches of wild habitat – like roads for animals! They provide a safe way to get from one place to another and allow species to feed, mate and recolonise habitats after local extinctions.

 ONLINE

Research the impact of fragmentation and habitat corridors on tiger populations using the links at www.brightredbooks.net

 ## THINGS TO DO AND THINK ABOUT

1 Explain how edge species may affect interior species as habitat fragments become smaller in size.

2 a Explain what is meant by a habitat corridor.

  b Explain how a habitat corridor can lead to an increase in biodiversity following a local extinction.

 ONLINE TEST

Head online and test yourself on this at www.brightredbooks.net

THREATS TO BIODIVERSITY 2

ONLINE

Watch the slideshow describing the top 10 invasive species in the UK at www.brightredbooks.net

ONLINE

Read about some of the problems caused by invasive species at www.brightredbooks.net

ONLINE

Read about the impact of climate change on biodiversity at www.brightredbooks.net

DON'T FORGET

Even a small change in the temperature of the atmosphere can have a significant effect on our climate.

INTRODUCED, NATURALISED AND INVASIVE SPECIES

An **indigenous** population is one which is native to a particular ecosystem and is present naturally without human intervention. An **introduced** species is a non-native species that has arrived in a new geographical area due to human activity, either intentionally or accidentally. When a non-native species spreads within wild communities, becomes established and can maintain its population through reproduction, it is known as a **naturalised species**. **Invasive species** are naturalised species which can spread rapidly and eliminate native species. They can have a negative impact in their new environment, since the absence of their usual predators, parasites, pests, diseases and competitors enables them to flourish and so outcompete native species for various resources. They may also prey on native species or hybridise with them. In Scotland, invasive species including giant hogweed and rhododendron are a threat to our native biodiversity. The grey squirrel, which was introduced to Britain in the nineteenth century, competes more successfully for both food and habitats and has caused a decline in the number of native red squirrels.

CLIMATE CHANGE AND ITS IMPACT ON BIODIVERSITY

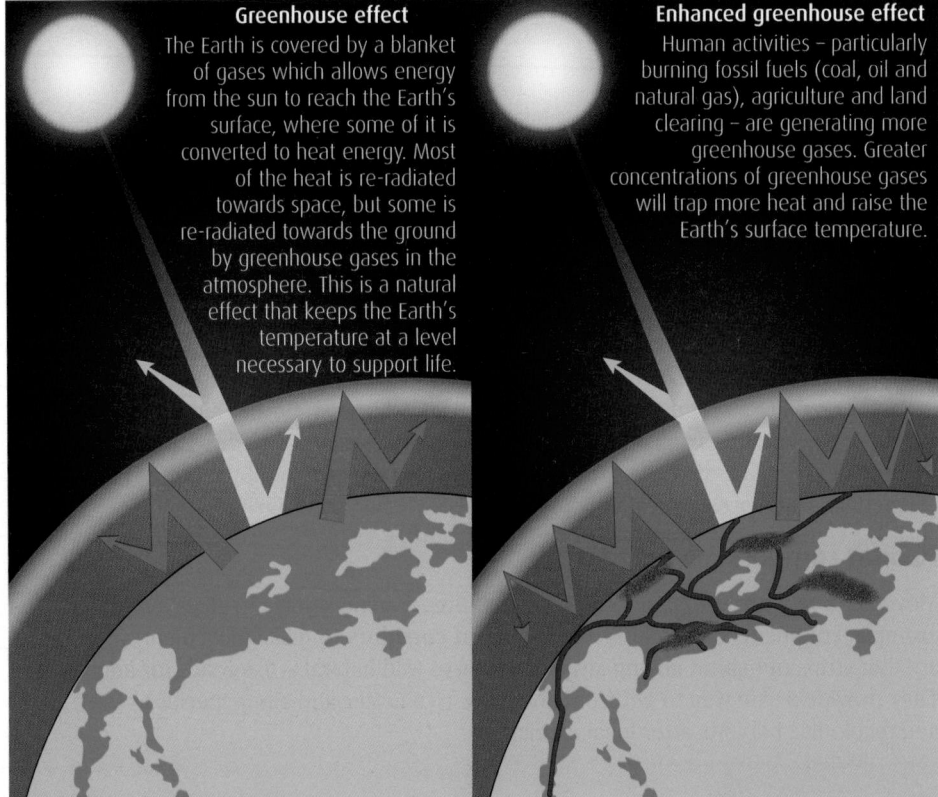

Greenhouse effect
The Earth is covered by a blanket of gases which allows energy from the sun to reach the Earth's surface, where some of it is converted to heat energy. Most of the heat is re-radiated towards space, but some is re-radiated towards the ground by greenhouse gases in the atmosphere. This is a natural effect that keeps the Earth's temperature at a level necessary to support life.

Enhanced greenhouse effect
Human activities – particularly burning fossil fuels (coal, oil and natural gas), agriculture and land clearing – are generating more greenhouse gases. Greater concentrations of greenhouse gases will trap more heat and raise the Earth's surface temperature.

Greenhouse gases, such as carbon dioxide and methane, absorb and reradiate the warmth of the sun. However, burning fossil fuels, farming and other human activities are increasing the levels of these gases in the atmosphere, enhancing the greenhouse effect. This contributes to global warming and is responsible for dramatic climatic changes which affect characteristics of ecosystems, leading to a decrease in biodiversity. It has been predicted that climate changes due to human activity could lead to a mass extinction event in the not too distant future. Some of the effects of climate change on biodiversity are shown in the table.

contd

Effect of climate change	Example/explanation
Change in distribution and abundance of species	Increased temperatures could force certain species to move to higher latitudes, where temperatures are more favourable to their survival.
Change in timing of seasonal events	Changes in the timing of key stages in the lifecycle of certain species, such as migration. Migrant species may arrive at a destination when food is not available.
Changes in composition of plant and animal communities	Melting of ice in the Arctic can cause a decline in the abundance of ice algae. This can result in a decline in the numbers of other organisms in an Arctic food web, such as plankton, cod, seals and polar bears.
Habitat loss	Changes in temperature and rainfall mean that some plants can no longer survive where they are currently growing. Many animals, including insects, rely on plants as their habitat. The loss of plant species will result in loss of these habitats and, so, the loss of animal species.
Migratory species leave for breeding areas earlier in the year	Increased competition with resident species.
Increased sea temperatures	Even a small rise in temperature (of 1°C) causes zooxanthellae algae to be ejected from coral polyps, which lose their colour. This is coral bleaching.

It is important for scientists to be able to predict the impact of climate change on biodiversity in the future so that strategies can be developed for the conservation of species. Computer models can simulate the climate and predict temperature, rainfall, extreme weather events and their effects on the distribution of species. These models can only speculate future events and may not be completely accurate but the real challenge is to find ways of reducing the generation of greenhouse gases.

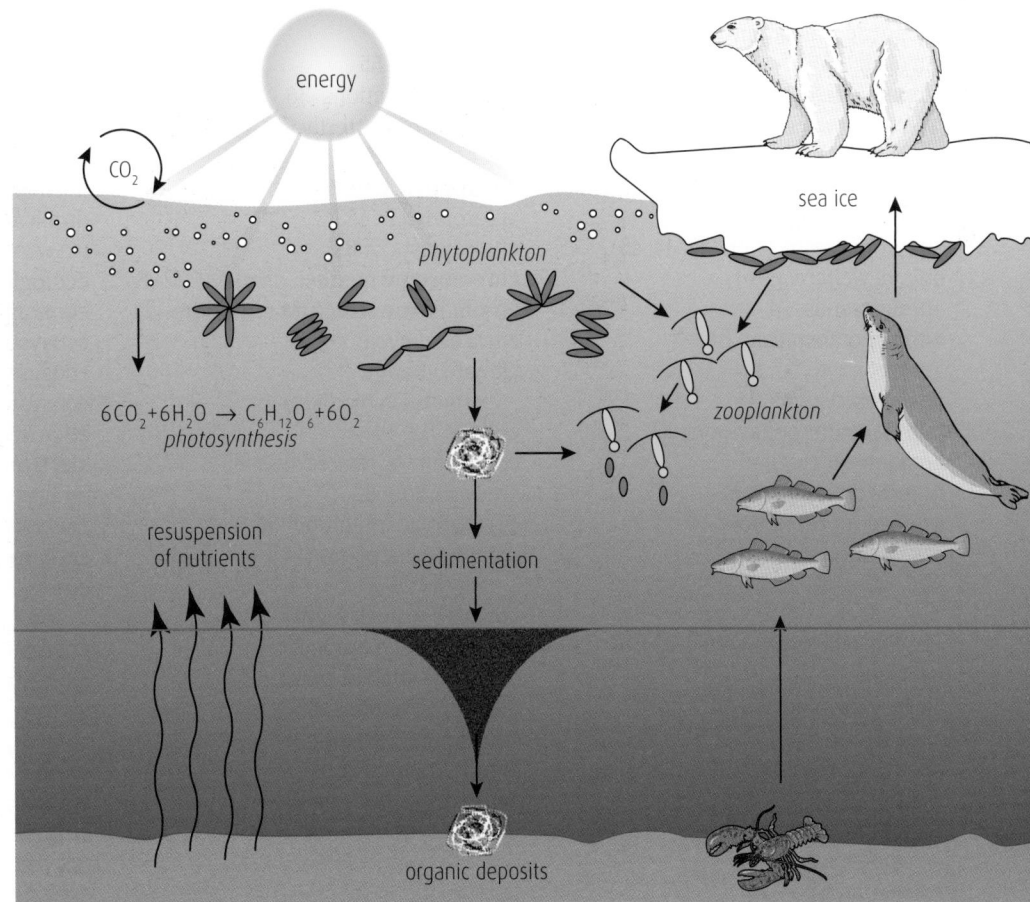

energy

CO_2

phytoplankton

sea ice

$6CO_2 + 6H_2O \rightarrow C_6H_{12}O_6 + 6O_2$
photosynthesis

zooplankton

resuspension of nutrients

sedimentation

organic deposits

Arctic food web. The loss of sea ice will affect all organisms.

THINGS TO DO AND THINK ABOUT

1 Which term is used to describe a naturalised species which eliminates native species?

2 A type of flatworm was introduced to Scotland from New Zealand in the 1960s. Its population has increased more quickly than it would have in its native habitat. Explain why this happened.

ONLINE TEST

How well have you learned this topic? Take the test at www.brightredbooks.net

INDEX

INDEX

nematodes 75, 77
neutral mutations 21, 29
non-competitive inhibition 38
nonsense mutations 20, 21
nucleotides 6

organelles 6–7, 34
outbreeding 70
overexploitation 90
oxygen delivery 46–7

parasites 78, 82
parasitic transmission 82
parasitism 82
pecking order 84
peppered moths 27
perennial weeds 74
pesticides 76, 77
pests 75
pharmacogenetics 33
phenotypes 71
phosphorylation 40
photosynthesis 62, 63, 64
    Calvin cycle 66–7
    light dependent stage 66
    limiting factors 67
phylogenetics 32
Phytophthora infestans 69
Phytoseiulus 78
plant field trials 68–9
plant growth
    limiting factors 63
    and productivity 64–7
plant productivity 67, 74–5
plasmids 60–1
Plasmodium 82
pluripotent cells 17
point mutations 20–1
pollination 87
polygenic inheritance 68
polymerase chain reaction (PCR) 9
polypeptide chains 6, 10, 14, 15
polypeptides 10
polyploidy 22–3
population bottlenecks 90
population growth 58
population growth rate, drawing
  graphs
   of 59
post-translational modification 15
potatoes 63, 69, 77, 89
predictive dormancy 53
primates, social behaviour 87
primers 9
prokaryotes 6, 34, 56, 83
    horizontal transfer in 24–5
protein pores 34
protein pumps 35

proteins
    fibrous 10
    globular 10
    membrane 34–5
    as respiratory substrates 45
    structure 10
    synthesis 12–15
protozoa 56

random mutations 26, 30
rates of reaction 38
reciprocal altruism 85
recombinant DNA technology 60–1
recombinant plasmids 60–1
red spider mites, control of 78
regulators, metabolism in 50–1
regulatory sequence mutations 20
relatedness 85
reptiles
    cardiovascular system 46
    lungs 47
resazurin dye 41
respiration 42–5
respiratory substrates 40, 45
restriction endonucleases 61
ribonucleic acid *see* RNA
ribosomal RNA (rRNA) 12
rice, genetically engineered 73
ritualistic behaviours 87
RNA 12–13
RNA polymerase 13, 36
RNA splicing 13
rRNA 12

selection 26–7
selective breeding 60, 70
selective insecticides 77
selective pesticides 77
sexual reproduction 24
sexual selection 26
silent mutations 21
single gene inheritance 68
skin grafts 19
social behaviour 84–7
social (dominance) hierarchies 84, 87
social organisms 84
sodium–potassium pump 35
solid growth media 57
specialised cells 16–17
speciation 30–1, 88
species, databases of 88
species diversity 89
spectrum, visible light 64
splice site mutations 20
stabilising selection 26
stationary phase 58
stem cells 17
    ethical issues 19
    research 18
    sources of 18
    therapeutic value 19

stereotypies 80
sterility 57
substitution 20
substrate concentration 38
sugar-phosphate backbone 6
survival, in adverse conditions 52–5
symbiosis 82
sympatric speciation 30–1
systemic fungicides 77
systemic insecticides 77
systemic pesticides 77

tapeworms 82
TAQ polymerase 55
temperature change, responses to 51
temperature regulation 51
test crosses 71
thymine (T) 6, 12
torpor, daily 53
total cell counts 58
totipotent cells 16
transcription 12–13
transfer RNA (tRNA) 12
transition state 36
translation 14
translocation 21
tRNA 12

unspecialised cells 16–17
uracil (U) 12

variable cell counts 58
vectors 60, 75, 82
vertical inheritance 24
viruses to eukaryotes, by horizontal
    transfer 24–5

weed killers 73, 76, 77
weeds 74
    chemical control 77
    cultural control 76
    prevention 76
whiteflies, control of 78

xanthophyll 65

yeast dehydrogenase investigation 41
yields 67

zooxanthellae algae 83